VOLUME ONE HUNDRED AND FIFTY FIVE

Progress in
MOLECULAR BIOLOGY AND TRANSLATIONAL SCIENCE

Metabolic Aspects of Aging

VOLUME ONE HUNDRED AND FIFTY FIVE

Progress in Molecular Biology and Translational Science

Metabolic Aspects of Aging

Edited by

MARY ANN OTTINGER
University of Houston, Houston, TX, United States

ELSEVIER

ACADEMIC PRESS
An imprint of Elsevier

Academic Press is an imprint of Elsevier
125 London Wall, London EC2Y 5AS, United Kingdom
525 B Street, Suite 1800, San Diego, CA 92101-4495, United States
50 Hampshire Street, 5th Floor, Cambridge, MA 02139, United States
The Boulevard, Langford Lane, Kidlington, Oxford OX5 1GB, United Kingdom

First edition 2018

ISBN: 978-0-12-809391-7
ISSN: 1877-1173

For information on all Academic Press publications
visit our website at https://www.elsevier.com/books-and-journals

Publisher: Zoe Kruze
Acquisition Editor: Ashlie M. Jackman
Editorial Project Manager: Joanna Collett
Production Project Manager: James Selvam
Cover Designer: Miles Hitchen

Typeset by Thomson Digital

CONTENTS

CONTRIBUTORS

Andrzej Bartke
Southern Illinois University School of Medicine, Springfield, IL, United States

R. Alexander Bentley
University of Tennessee, Knoxville, TN, United States

Andre Catic
Huffington Center on Aging, Stem Cells and Regenerative Medicine Center, Dan L. Duncan Comprehensive Cancer Center, Baylor College of Medicine, Houston, TX, United States

Justin Darcy
Southern Illinois University School of Medicine, Springfield, IL, United States

Yih-Woei Fridell
National Institute on Aging, NIH, Bethesda, MD, United States

Deborah G. Hamilton
Texas Obesity Research Center and Health and Human Performance, University of Houston, Houston, TX, United States

Marc T. Hamilton
Texas Obesity Research Center and Health and Human Performance, University of Houston, Houston, TX, United States

Michal M. Masternak
University of Central Florida, Orlando, FL, United States;
The Greater Poland Cancer Centre, Poznan, Poland

Michael J. O'Brien
Texas A&M University–San Antonio, San Antonio, TX, United States

Mary Ann Ottinger
University of Houston, Houston, TX, United States

Pamela Anne Quiroz
Center for Mexican American Studies, University of Houston, Houston, TX, United States

Corinna N. Ross
Texas A&M University–San Antonio, San Antonio, TX, United States

Felipe Sierra
National Institute on Aging, NIH, Bethesda, MD, United States

Stephen J. Spann
University of Houston, Houston, TX, United States

Berta Victoria
University of Central Florida, Orlando, FL, United States

Theodore W. Zderic
Texas Obesity Research Center and Health and Human Performance, University of Houston, Houston, TX, United States

PREFACE

The process of aging is multidimensional with dynamic changes in physiological systems accompanied by progressive dysregulation of interlocking rhythmic systems ultimately leading to declining function. Lifespan in humans has been progressively increasing, and there is a rapid growing literature documenting aging processes both in healthy aging individuals and with diseases that have increased incidence, such as neurodegenerative diseases. Community, culture, lifestyle, and accessibility to healthcare are powerful factors that greatly impact longevity and healthy aging. Insights are gained from the tremendous variability in the timing, pattern, and chronology of aging across vertebrate species. Further, the rate and characteristics of aging across vertebrate species vary greatly, with notable adaptations in long-lived species. These adaptations have been studied to understand some of the underlying and conserved mechanisms involved in the aging process. Additionally, the adaptive mechanisms utilized by long-lived and hibernating species provide insights into adaptive mechanisms that may also be conserved across vertebrates or even invertebrates. This volume has been organized to reflect many of the dimensions inherent in aging, with metabolism being one of the foremost issues occurring in today's society. Furthermore, obesity has emerged as a major problem in the United States and worldwide. Chapters in this volume consider aging, metabolism, and culture as interactive constituents influencing genetic makeup to culminate in individual health and lifespan. Community framework, support, culture, and education are powerful factors in longevity and healthy aging. There are many aspects denoting healthy aging with the ultimate goal of extending the highest quality of life throughout the greatest proportion of the lifespan. Although individuals are unique in their aging process, education and lifestyle can promote health and delay the onset and progression of common ailments that occur during aging. Age-related changes are also affected by culture, as well as social structures. As mentioned earlier, individual lifestyle exerts tremendous impacts on the process of aging. As more of our working life become sedentary, the impacts of exercise or the lack of activity because major factors in overall metabolic function, which in turn impacts longevity and healthy aging. Again, community, culture, and education become critical factors as a major segment of our populations become aged.

Understanding the integrative and coordinated aspects of our physiological systems is critical to understanding aging processes. A myriad of molecular and cellular level changes accompany the functional alterations in metabolism during aging. The age-related changes in the function of endocrine, neuroendocrine, and physiological systems impact neural, immune, metabolic, skeletal systems resulting in impaired overall physiological and behavioral responses and resilience. Metabolic processes are central to the physiology of aging and selected processes are also conserved across a number of vertebrates. Finally, there is a "fountain of youth" that provides an intervention for the age-related decline that occurs in metabolic function and associated physiological systems. Calorie restriction, exercise, and treatments that ameliorate diabetic and other conditions have proven to be effective interventions. Further, development of reliable biomarkers useful in predictive and diagnostic medicine will provide powerful tools for preventive medicine and individual adjustments in lifestyle favorable to healthy aging. The contributions of authors to this volume have provided an excellent overview of these complexes and interwoven physiological processes.

CHAPTER ONE

Longevity, Metabolic Disease, and Community Health

Stephen J. Spann[1], Mary Ann Ottinger
University of Houston, Houston, TX, United States
[1] Corresponding author. E-mail address: maotting@umd.edu

Contents

Abstract

In the United States, the average lifespan has increased despite the dramatic increase in obesity, diabetes, and other conditions that worsen during aging. As the longevity of US population increases, it is critical to understand the factors that impact aging populations especially as age-related disease and declining health becomes more prevalent. Diabetes related to obesity has become much more prevalent throughout the United States and globally. Further, the prevalence of age-related health problems accelerate in lower income communities with less access to health care. All these factors become critical as individuals age. Furthermore, in communities with less availability to health care, diagnosis may be delayed and treatments are initiated at a much later stage in disease. As such, the costs of medical care skyrocket leading to higher costs both to the community and to taxpayers. This chapter reviews some key health problems and issues in community health and healthy aging, recognizing the importance of organizations and programs that provide education and support to the aging population. Finally, cultural differences in approaches to healthy aging provide important insights and lessons for optimizing quality of life during aging.

Progress in Molecular Biology and Translational Science, Volume 155
ISSN 1877-1173
http://dx.doi.org/10.1016/bs.pmbts.2017.11.015

1. INTRODUCTION

Average lifespan has dramatically increased over the past 100 years with biomedical advances, antibiotics, immunizations, and other lifesaving technologies. However, modern lifestyle has been associated with reduced physical activity and a growing trend for obesity and accompanying health problems, such as diabetes. It has long been recognized that there are regional, national, and global differences in the prevalence of selected diseases, as well as in extended lifespan. Furthermore, it is critical to understand basic biological mechanisms operating during the process of aging, especially those that are conserved which may provide the basis for treatments and interventions effective across a range of individuals.[1] Moreover, the role of community is increasingly recognized as a central and critical component of the longevity and health of aging individuals. This chapter examines population trends and how the community can contribute to overall healthy aging for both individuals and the populations as a whole. Selected aspects of cultural differences in lifespan and aging are also considered.

2. LONGEVITY AND POPULATION TRENDS

Many of these statistics have been tracked for a number of years and show fascinating projections in the ages of populations for the United States and other countries. The United Nations Department of Economic and Social Affairs, Population Division has provided excellent historical data and probabilistic projections that show trends in population from 1950 to 2100[2] (https://esa.un.org/unpd/wpp/Graphs/Probabilistic/ODR/70plus/25-69/). In the case of US population, the number of individuals over 65 years of age has been steadily increasing with a sharper increase between 2010 and 2030 (Fig. 1). Probabilistic projections show an array of curves, all of which continue to a relatively steep increase shown by the median and all other projections as we approach the year 2100. Comparing these curves to those for projected population increase shows a consistent relative increase in the aging population that mirrors the overall growth in the US population (Fig. 2). Fig. 3 shows probabilistic changes in age and population of the United States through the year 2100, with continued population increase for individuals over 65 years of age. When the potential for old-age dependency for the population over 70 years of age is projected, there is a predicted

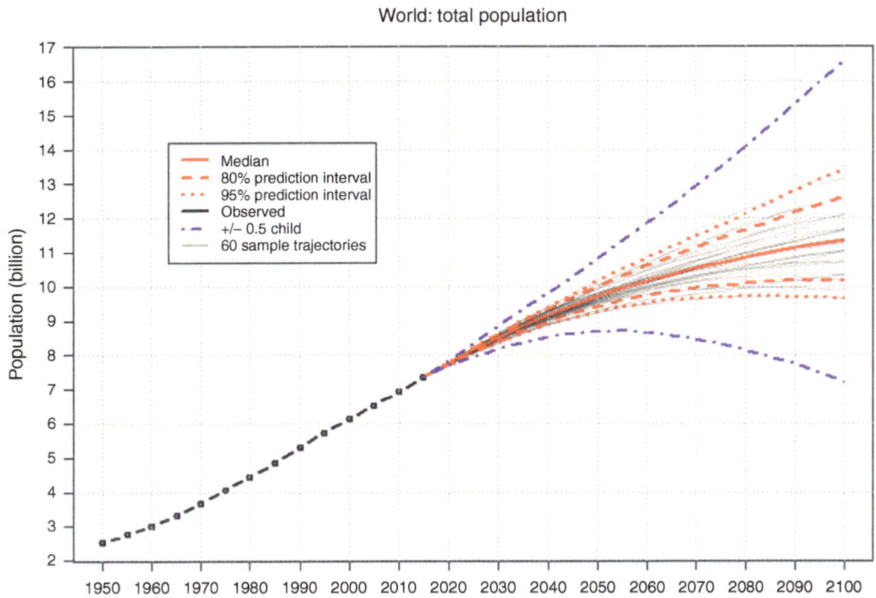

Fig. 1 The estimated and probabilistic projections for the total world population uses projections of total fertility and life expectancy at birth, based on the 2015 World Population Prospects with a Bayesian Hierarchical Model. Probabilistic median and confidence intervals show high and low variant (+/– 0.5 child) of the 2015 Revision of the World Population Prospects. *From United Nations Department of Economic and Social Affairs, Population Division, 2015. Word Population prospects: The 2015 Revision. Available from: http://esa.un.org/unpd/wpp/; https://esa.un.org/unpd/wpp/Graphs/DemographicProfiles/; https://esa.un.org/unpd/wpp/Graphs/Probabilistic/POP/TOT/).*

increase in dependency that mirrors the rising number of aging individuals associated with increased lifespan. An in-depth report by the National Institute on Aging of the National Institutes of Health and The World Health Organization considered current challenges faced by aging individuals.[3] Some of the challenges articulated include the rising incidence of dementia in individuals aged 85 and older. Additionally, family structure to care for aging individuals is complicated by fewer family members to care for aged parents and relatives. As life expectancy increases, the potential for age-related disease rises; this is complicated by the increasing frequency of obesity and diabetes that impacts health and aging. It is concerning that the prevalence of chronic disease persists at a relatively high level in the United States, even as compared to Europe. Further, more effective interventions are being developed as more is understood about the essential metabolic pathways

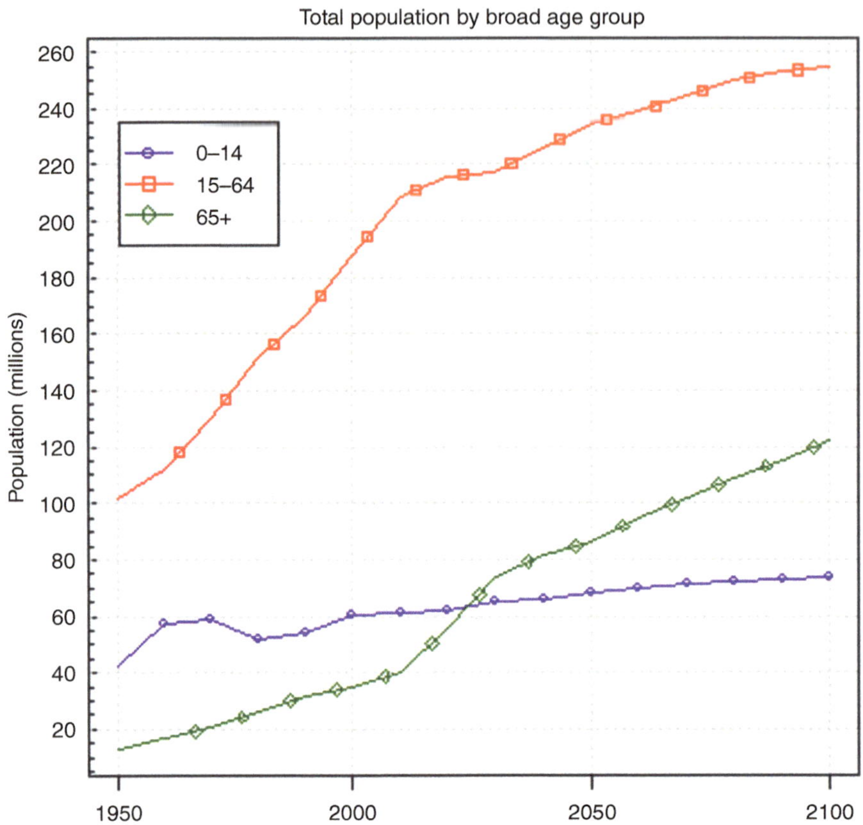

Fig. 2 Demographic profiles for the United States with estimates for future increases. *From United Nations Department of Economic and Social Affairs, Population Division, 2015. Available from: https://esa.un.org/unpd/wpp/Graphs/DemographicProfiles/.*

involved in extending the healthspan.[4] Dementia is recognized as a costly debilitating disease that has been increasing as the age of the population increases, with relatively high prevalence in less developed countries. In summary, it is critical to maintain individual health over the lifespan to optimize healthy aging as well as to reduce burden to families, both economic and emotional.

3. OBESITY, METABOLIC SYNDROME, AND AGING

Metabolic syndrome encompasses many conditions and associated diseases arising from genetics and lifestyle. Obesity is regarded as a highly

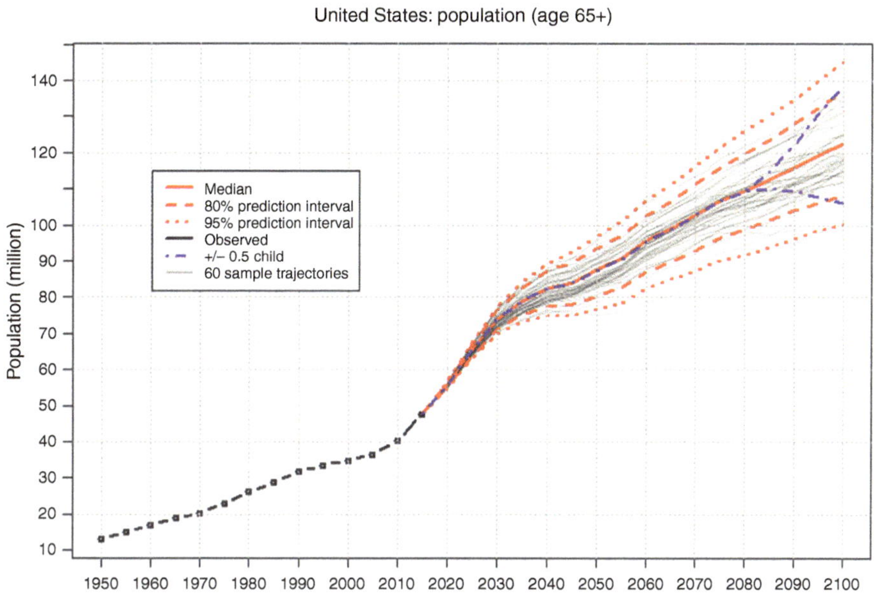

Fig. 3 Probabilistic population projections of total fertility and life expectancy at birth, based on the 2015 Revision of the World Population Prospects, using a Bayesian Hierarchical Model. Median, 80% and 95% prediction intervals shown with variation (+/− 0.5 child). *From United Nations Department of Economic and Social Affairs, Population Division, 2015. Word Population Prospects: The 2015 Revision. Available from: http://esa.un.org/unpd/wpp/; https://esa.un.org/unpd/wpp/Graphs/DemographicProfiles/; https://esa.un.org/unpd/wpp/Graphs/Probabilistic/POP/70plus/.*

prevalent risk factor, associated with a higher incidence of a number of diseases and conditions including heart disease, and some types of dementia and cancer. During aging, regulation of glucose metabolism deteriorates; necessitating dietary and sometimes pharmacological support to improve health.[5] Further, inflammation has been another key component of aging processes; nutritional, as well as environmental factors may help to ameliorate or conversely exacerbate these inflammatory processes.[6,7] Many studies have pointed to the critical role of the mitochondria and alterations associated with increased oxidative damage and diminished efficiency in energy production and utilization.[8] As such, mitochondrial stress would be a key component of this pathway.[9] This pathway then becomes a critical driver in the cascade of age-related changes encompassing protein homeostasis, nutritional utilization, adipose tissue responses, and a number of associated metabolic pathways.[10]

The question of heritability of healthy aging has received a great deal of global attention, with multiple longitudinal studies of both individual and population level trends in humans. An inherent issue in all of these studies is the assessment of the relative influence of genetic predisposition and environmental factors as drivers for longevity and healthy aging. A study in 421 long-lived families revealed that genetic variants modulating glucose, insulin, and triglyceride variation were associated with familial longevity, providing potential biomarkers of healthy aging.[11] Some estimate that 25% of the variation in human longevity is due to genetics; however, epigenetic and other environmentally induced alterations can modify the aging process and longevity.[12] Not surprisingly, there are conflicting data from a study of octogenarians compared to centenarians, which did not reveal distinct genetic factors.[13] Data collected specifically from individuals with exceptional longevity have revealed some common characteristics, including compression and later onset of morbidity and some potential variants identified in this population.[14,15] Further, attempting to capture the metabolic profile for a longevity phenotype must include signaling pathways involved in immune and antioxidant responses, as well as the complexities of monitoring nutritional factors modulating these responses.[16] Answers emerging from these studies are not conclusive, but provide intriguing data that support a complex interaction between expression of genomic bases and environmental stressors. These studies also clearly show the critical importance of community and culture acting to promote the health and longevity of individuals within varied populations. Linking economics to longevity has pointed to significant issues of diet and nutrition, timeliness of health care and access to clinics, services, transportation, and other factors that enhance life-style for elderly individuals.

Caloric restriction (CR) and other nutritional treatments have been extensively studied across a range of animal models and some trials have been conducted with human subjects. There is a very rich literature on studies about the effects of CR in animal models, which also shows benefits of CR on aging processes; however, the responses are complex and physiological responses to CR are not uniformly transferable across classes of organisms.[17] Not surprisingly, adipose tissue is dynamic and quickly adapts to nutritional status through mitrochondrial and other cellular responses.[18] As such, beneficial effects of CR are expressed on metabolic function, with biomarkers of health and metabolic function showing improvement, as well as slowed aging in the realms of inflammatory markers, adiposity, and endocrine measures related to metabolism.[19] Dietary modifications including monitoring of overall calories and intermittent fasting have also proven beneficial to overall health.[20]

4. COMMUNITY HEALTH AND HEALTHY AGING

The intersection of community health and healthy aging is recognized as critical for quality of life, longevity, and healthy aging. As discussed earlier, United States population is aging and with aging comes an increased risk for age-related diseases. The incidence of Alzheimer's disease (AD) as well as other forms of dementia is increasing, coincident with the aging population. Further, there are data indicating a higher prevalence of AD in the Latino community.[21] These and other data strongly support the critical role of the community, with models of aging-friendly communities gaining attention.[22] The advent of many forms of digital technologies provides options for individuals to continue to live more independently.

5. SOCIAL DETERMINANTS OF HEALTH AND AGING

There is a growing recognition of the importance of the impact of social determinants (social and economic conditions) on individual health status. Social determinants can be defined as the circumstances in which people are born, live, work, and age, as well as the systems set up to address illness.[23] Categories of social determinants include social and community context, education, economic stability, neighborhood and built environment, and health and healthcare. There is a growing body of evidence that confirms the impact of social determinants on the development and progression of chronic diseases, such as diabetes mellitus.[24–26] Social support, a perception that one is accepted, cared for, and provided with assistance from certain individuals or a specific group or the realization of actual support received from another and can be considered a social determinant of health, has been shown to have an important impact on health outcomes in patients with chronic diseases, such as diabetes mellitus,[26a] and on physical activity in older adults.[27] Social determinants play an important role in the health of communities, and in the aging of their populations.

6. CULTURE AND AGING

Culture shapes the way in which individuals age by influencing all of the determinants of aging. Cultural values and traditions shape how a society views older people and the aging process. Culture affects whether or not coresidency with younger generations in multigenerational households is the preferred way of living, cultural factors affect health behaviors and lifestyles, including diet, physical activity, and use of other noxious substances.[28]

REFERENCES

1. Kaeberlein M, Rabinovitch PS, Martin GM. Healthy aging: the ultimate preventative medicine. *Science.* 2015;350(6265):1191–1193.
2. United Nations. Department of Economic and Social Affairs, Population Division 2015. United States of America: Populations (Age 65+) in World Population Prospects: The 2015 Revision. Available from: http://esa.un.org/unpd/wpp/
3. Suzman R, Beard J. Global Health and Aging. National Institute on Aging, National Institutes of Health, U.S. Department of Health and Human Services; World Health Organization. 2011;1–23
4. Longo VD, Antebi A, Bartke A, Barzilai N, Brown-Borg HM, Caruso C, Curiel TJ, de Cabo R, Franceschi C, Gems D, Ingram DK, Johnson TE, Kennedy BK, Kenyon C, Klein S, Kopchick JJ, Lepperdinger G, Madeo F, Mirisola MG, Mitchell JR, Passarino G, Rudolph KL, Sedivy JM, Shadel GS, Sinclair DA, Spindler SR, Suh Y, Vijg J, Vinciguerra M, Fontana L. Interventions to slow aging in humans: Are we ready? *Aging Cell.* 2015; 14(4):497–510.
5. Brewer RA, Gibbs VK, Smith Jr DL. Targeting glucose metabolism for healthy aging. *Nutr Healthy Aging.* 2016;4(1):31–46.
6. Pickering AM, Lehr M, Miller RA. Lifespan of mice and primates correlates with. immunoproteasome expression. *J Clin Invest.* 2015;125(5):2059–2068.
7. Szarc vel Szic K, Declerck K, Vidaković M, Vanden Berghe W. From inflammaging to healthy aging by dietary lifestyle choices: is epigenetics the key to personalized nutrition? *Clin Epigenetics.* 2015;7:33.
8. Gonzalez-Freire M, de Cabo R, Bernier M, Sollott SJ, Fabbri E, Navas P, Ferrucci L. Reconsidering the role of mitochondria in aging. *J Gerontol A Biol Sci Med Sci.* 2015; 70(11):1334–1342.
9. Hill S, Van Remmen H. Mitochondrial stress signaling in longevity: a new role for. mitochondrial function in aging. *Redox Biol.* 2014;2:936–944.
10. Carmona JJ, Michan S. Biology of healthy aging and longevity. *Rev Invest Clin.* 2016; 68(1):7–16.
11. Deelen J, van den Akker EB, Trompet S, van Heemst D, Mooijaart SP, Slagboom PE, Beekman M. Employing biomarkers of healthy ageing for leveraging genetic. studies into human longevity. *Exp Gerontol.* 2016;82:166–174.
12. Passarino G, De Rango F, Montesanto A. Human longevity: Genetics or Lifestyle? It takes two to tango. *Immun Ageing.* 2016;13:12.
13. Nickles D, Madireddy L, Patel N, Isobe N, Miller BL, Baranzini SE, Kramer JH, Oksenberg JR. Whole genome sequences of 2 octogenarians with sustained cognitive abilities. *Neurobiol Aging.* 2015;36(3):1435–1438.

14. Ismail K, Nussbaum L, Sebastiani P, Andersen S, Perls T, Barzilai N, Milman S. Compression of morbidity is observed across cohorts with exceptional longevity. *J Am Geriatr Soc.* 2016;64(8):1583–1591.
15. Sebastiani P, Gurinovich A, Bae H, Andersen S, Malovini A, Atzmon G, Villa F, Kraja AT, Ben-Avraham D, Barzilai N, Puca A, Perls TT. Four genome-wide association studies identify new extreme longevity variants. *J Gerontol A Biol Sci Med Sci.* 2017;72(11):1453–1464.
16. Dato S, Bellizzi D, Rose G, Passarino G. The impact of nutrients on the aging rate: a complex interaction of demographic, environmental and genetic factors. *Mech Ageing Dev.* 2016;154:49–61.
17. Mitchell SJ, Madrigal-Matute J, Scheibye-Knudsen M, Fang E, Aon M, González-Reyes JA, Cortassa S, Kaushik S, Gonzalez-Freire M, Patel B, Wahl D, Ali A, Calvo-Rubio M, Burón MI, Guiterrez V, Ward TM, Palacios HH, Cai H, Frederick DW, Hine C, Broeskamp F, Habering L, Dawson J, Beasley TM, Wan J, Ikeno Y, Hubbard G, Becker KG, Zhang Y, Bohr VA, Longo DL, Navas P, Ferrucci L, Sinclair DA, Cohen P, Egan JM, Mitchell JR, Baur JA, Allison DB, Anson RM, Villalba JM, Madeo F, Cuervo AM, Pearson KJ, Ingram DK, Bernier M, de Cabo R. Effects of sex, strain, and energy intake on hallmarks of aging in mice. *Cell Metab.* 2016;23(6):1093–1112.
18. Lettieri Barbato D, Aquilano K. Feast and famine: adipose tissue adaptations for healthy aging. *Ageing Res Rev.* 2016;28:85–93.
19. Lettieri-Barbato D, Giovannetti E, Aquilano K. Effects of dietary restriction on adipose mass and biomarkers of healthy aging in human. *Aging.* 2016;8(12):3341–3355.
20. Fontana L, Partridge L. Promoting health and longevity through diet: from model organisms to humans. *Cell.* 2015;161(1):106–118.
21. Vega IE, Cabrera LY, Wygant CM, Velez-Ortiz D, Counts SE. Alzheimer's disease in the Latino community: intersection of genetics and social determinants of health. *J Alzheimers Dis.* 2017;58(4):979–992.
22. Scharlach AE. Aging in context: individual and environmental pathways to aging-friendly communities-The 2015 Matthew A. Pollack Award Lecture. *Gerontologist.* 2017;57(4):606–618.
23. Healthy People. Secretary's Advisory Committee on Health Promotion and Disease Prevention Objectives for 2020. Healthy People 2020: An Opportunity to Address the Societal Determinants of Health in the United States. Available from: http://www.healthypeople.gov/2020/topicsobjectives2020/overview.aspx?topicid=39
24. Walker RJ, Gebregziabher M, Martin-Harris B, Egede LE. Independent effects of socioeconomic and psychological social determinants of health on self-care and outcomes in type 2 diabetes. *Gen Hosp Psychiatry.* 2014;36(6):662–668.
25. Walker RJ, Smalls BL, Campbell JA, Strom JL, Egede LE. Impact of social determinants of health on outcomes for type 2 diabetes: a systematic review. *Endocrine.* 2014;47:29–48.
26. Walker RJ, Strom J, Egede LE. Impact of race/ethnicity and social determinants of health on diabetes outcomes. *Am J Med Sci.* 2016;351(4):366–373.
26a. Strong JL, Egede LE. The impact of social support on outcomes in adult patients with type 2 diabetes: a systematic review. *Curr Diab Rep.* 2012;12(6):769–781.
27. Smith GL, Banting L, Eime R, O'Sullivan G, van Uffelen JGZ. The association between social support and physical activity in older adults: a systematic review. *Int J Behav Nutr Phys Act.* 2017;14(1):56.
28. WHO. Active ageing, a policy framework. WHO/NMH/NPH/02.8. Available from: http://www.who.int/ageing/publications/active_ageing/en/

FURTHER READING

Kim KI, Gollamudi SS, Steinhubl S. Digital technology to enable aging in place. *Exp Gerontol.* 2017;88:25–31.

CHAPTER TWO

Metabolic Aspects of Aging

Yih-Woei Fridell[1], Felipe Sierra
National Institute on Aging, NIH, Bethesda, MD, United States
[1] Corresponding author. E-mail address: yih-woei.fridell@nih.gov

Contents

Abstract

Metabolic interventions involving undernutrition but not malnutrition (e.g., caloric restriction, CR) are effective strategies for improving both health and longevity in species ranging from lower organisms to nonhuman primates. Initial human trials to test the effects of sustained, reduced energy intake have yielded promising health benefits. Through intense research efforts in understanding the molecular mechanisms of CR, three cellular pathways have now been identified although the precise details remain unknown. More recently, circadian regulation has been recognized as a novel mediator for CR effects in mice. Harnessing the molecular insights into CR, novel nutritional interventions and pharmacological application of CR mimetics have been tested showing great promise in simultaneously improving metabolic function and providing overall health benefits. Additional research is needed to identify efficacious therapeutics that can be safely and practically translated to human studies in promoting healthspan.

1. INTRODUCTION

Aging and metabolism are tightly linked. More than 80 years ago, McCay *etal.* first demonstrated that reduced food intake without malnutrition

Progress in Molecular Biology and Translational Science, Volume 155
ISSN 1877-1173
http://dx.doi.org/10.1016/bs.pmbts.2017.12.015

extends both mean and maximal lifespan in rats as compared to ad libitum fed controls.[1] Following this initial finding however, it was not until the 1980s that researchers in the aging field began in earnest to understand the impact of nutrition and metabolism on aging. Since then, a tremendous advance has been gained in delineating molecular elements that link metabolic homeostasis to the biology of aging where caloric restriction (CR), also referred to as dietary restriction (DR), remains a conserved and very robust environmental intervention to extend both lifespan and healthspan in diverse organisms.[2] The intense research effort leading to this conclusion served as foundation for the first human trial on reduced energy intake in 2007.

An enormous amount of literature has been accumulated over the last 3 decades in understanding the molecular mechanisms of CR and its translational potential as well as novel nutritional interventions mimicking the CR effects. In this article we have focused our attention on selected metabolic topics as they relate to aging, including an overview of CR and emerging mechanisms, the first human trial on sustained, reduced energy intake, novel nutritional strategies that improve healthspan and longevity without restricting all macronutrients, and finally, perspectives on metabolism-based aging interventions.

2. CR FROM YEAST TO PRIMATES

Reduced caloric intake has been demonstrated to extend lifespan in model systems ranging from yeast[3] to rodents,[4] and including the fruit fly *Drosophila*[5] and the nematode *Caenorhabditis elegans*.[6,7] In single cells or invertebrates where controlling ingested calories is not feasible, reduced caloric intake is achieved through dilution of macronutrients. For example, reduction of sugar content in yeast cultures is sufficient to extend the replicative lifespan of this organism,[8] whereas dilution of *E. coli*, the food source for worms, also leads to lifespan extension.[9] In fruit flies, reduced concentrations of sucrose or yeast, two major constituents in the diet providing carbohydrate and protein, respectively, are correlated with extended longevity.[10] In rodents, CR remains the most reproducible experimental intervention in extending lifespan by up to 50% as compared to ad libitum fed controls.[11] Interestingly, however, when CR was given to 41 different recombinant inbred strains of mice, strain-specific effects on lifespan were observed, suggesting gene–diet interactions in lifespan determination.[11] Importantly, in addition to extending longevity, those studies also demonstrated multiple beneficial metabolic changes associated with CR. For example, yeast cells under CR have increased respiration,[8] whereas altered mitochondrial energy metabolism was measured in worms under CR.[12] In *Drosophila*, a decrease in reactive oxygen species production and altered mitochondrial function were measured.[13] In rodents, long-term CR has a profound effect on white adipose tissue mass and its transcriptional profile.[14]

While two parallel CR studies in the nonhuman primate rhesus monkeys revealed conflicting results regarding lifespan effects, both studies demonstrated clear benefits in terms of metabolic fitness and other age-related parameters.[15–17] In terms of longevity, the study performed at the University of Wisconsin Madison (UW) reported health benefits of CR in age-related survival[15] as well as all-cause survival.[16] In contrast, the study performed at the National Institute on Aging (NIA) demonstrated no significant effect of CR on survival while noting health improvements in CR animals.[17] In an effort to explain these discrepant results, a comprehensive assessment of several aspects of the study design between the two sites has been recently reported.[18] The analysis reveals fundamental differences in the administration of CR to the two parallel cohorts of monkeys in terms of age of CR onset, macro- and micronutrient constituents, feeding timing, and frequencies that ultimately result in significant differences in body weight, food intake, and additional biometrics.[18] Given that CR diet commenced only in adult monkeys at the UW whereas CR diet was initiated in animals of a wider age spectrum at the NIA, the CR-associated improvements in survival at UW suggest that the onset of CR is critical in determining the extent of CR effects on survival at least in primates.[18] Nonetheless, it is important to note that a delay in age-related morbidities was observed in CR monkeys as compared to controls at both study sites. Specifically, cancer incidence is greatly reduced in both studies, whereas the decreased incidence of diabetes found only at NIA is consistent with the improved insulin resistance in monkeys at UW.[18] Taken together, despite intrinsic differences in study designs between the two study sites contributing to discrepant survival effects, long-term CR in nonhuman primates was shown to provide significant health benefits in delaying age-associated diseases, possibly predicting similar benefits in human health.

3. CALERIE STUDY

To begin to understand the effects of long-term CR in humans, the Comprehensive Assessment of Long-term Effects of Reducing Intake of

Energy (CALERIE) study marked the first multicenter, randomized controlled trial to test the CR paradigm in humans. The initial small-scale, phase I study was designed to establish safety and feasibility of CR in nonobese humans.[19–22] The CALERIE-2 study began in 2007 with the recruitment of over 200 healthy, nonobese adults who then underwent 2 years of sustained CR with a goal of 25% reduction in energy intake.[23] Despite the intended 25% reduction, when determined retrospectively by the intake-balance method at 6-month intervals, the average CR was at 11.9% over the 2-year period.[24–27] At this level of reduction in energy reduction, improvements in cardiometabolic risk factors and decreased circulating levels of T3 and TNF-α were measured, suggesting a positive impact on healthspan.[28] Changes in body composition during aging often lead to loss of fat-free mass (FFM) and increases in adiposity, or fat mass (FM) that can increase the risk of metabolic disorders. To examine short- and long-term effects of CR on body composition, researchers in the CALERIE-2 study have found that CR participants displayed a higher FFM and lower FM at 24 months than control subjects.[29] Interestingly, men in the CR group lost significantly more overall FM and FFM as a percentage of weight loss than women in the same group.[29] These sex-specific changes in body composition in response to CR are not fully understood and could potentially be the result of the smaller number of men than women participating in the study. In summary, although the lower than intended CR level achieved in the CALERIE-2 study has rendered the full physiological outcomes of sustained, 25% reduction in energy intake unattainable, this unique 2-year study of human CR has nonetheless demonstrated a great potential in improving healthspan parameters with a sustained reduction in energy intake.

4. FASTING AND FASTING MIMICKING DIET

Instead of sustained, reduced daily energy intake regimens during CR, extreme forms of energy deprivation such as fasting have been explored to develop effective strategies to improve health and delay aging. Intermittent fasting (IF) is achieved either through alternate day fasting or 2 days a week fasting in which individuals consume only water with intervening periods of normal food intake. Periodic fasting (PF) refers to IF with cycles of fasting lasting 3 days or longer.[30]

Fasting dramatically extends longevity in bacteria,[31] yeast,[32] and *C. elegans*,[33] whereas IF (alternate day fasting) applied to young mice improves lifespan.[34] The beneficial effects of IF go beyond longevity as IF in laboratory rodents has been shown to protect against a wide range of age-associated diseases.[35] In mammals, fasting greatly diminishes hepatic glycogen, a major supply source of cellular glucose. Consequently, nonhepatic glucose, ketone bodies produced in the liver and free fatty acids are utilized as energy sources through activation of alternative metabolic programs.[30] Similarly, PF promotes adaptive cellular stress response signaling pathways, enhanced mitochondrial function, DNA repair, autophagy, and stem cell-based regeneration as well as sustained metabolic effects.[35] Interestingly, in rodents, IF induces physiologically similar responses as those following regular aerobic exercises including insulin sensitivity and cardioprotective effects.[36–38]

Human studies have attributed the antiaging effects of fasting to changes in levels of IGF-1, IGFBP1, glucose, and insulin. It is well documented that as a major growth factor in mammals, IGF-1's level is negatively associated with longevity in epidemiological studies.[39] Therefore, it is of particular interest that 5 days of fasting causes a significant decrease in IGF-1 levels and increased concentrations of IGFBP1, a chief IGF-1 inhibiting protein.[40] To translate fasting and associated health benefits and standardize interventions, a low protein/low calorie, fasting mimicking diet (FMD) has been developed that elicits molecular changes mirroring those caused by fasting.[41] In the initial animal study, mice were fed with FMD for 4 days twice a month with ad libitum diet between FMD cycles. Following 4.5 months of FMD cycles, mice fed this diet had improved metabolism and cognitive function, increased regenerative potential, and extended lifespan when compared to controls.[41] Building upon these findings, human FMD (34%–54% of the normal caloric intake) was formulated with components that modulate levels of IGF-1, IGFBP-1, glucose, and insulin. Thirty-eight healthy adults were then randomized into two groups, one was given FMD for 5 days every month for 3 months (3 cycles) with normal diets between cycles, whereas the control group continued their normal diets. The analysis of biomarkers and risk factors for aging and chronic diseases such as CVD, diabetes, and cancer has revealed beneficial effects following three FMD cycles, suggesting a great potential in applying PF strategies to promote human healthspan.[41] Future studies enrolling a larger number of subjects will be necessary to confirm these results.

5. PROTEIN RESTRICTION

Are the health and aging benefits of CR attributable to reduced caloric intake or the composition of the macronutrients in the diet? There has been an intense effort in teasing out the specific effects of each of the three major dietary macronutrients: carbohydrate, protein, and fat on CR. Interestingly, in fruit flies, when total calories remain constant, reduction of yeast (protein source) has a much greater effect per calorie than does sugar in reducing mortality and extending lifespan.[42] In another study, the relationships between nutrients, longevity, and reproductive fitness were investigated extensively in a study where fruit flies were fed 28 different diets encompassing a wide range of protein/carbohydrate ratios.[43] Consistent with previous findings,[42] it was shown that low protein-to-carbohydrate ratios are beneficial for both longevity and reproductive fitness in terms of lifetime egg production and egg-laying rate.[43]

The notion that calorie intake is not the dominant factor in mediating the benefits of CR in insects has borne out in rodent studies.[44] Mice fed 1 of 25 diets with varying contents of protein, carbohydrate, and fat ad libitum have profoundly different responses to food intake, cardiometabolic phenotypes, and longevity.[44] Compensatory feeding occurs when the concentrations of either protein or carbohydrate decrease in the diet, whereas fat concentration has minimal effect on food intake. When compensatory feeding for protein was limited and replaced with carbohydrate to suppress protein intake, longevity and health were optimized. Among the wide range of diets tested, the greatest median lifespan was achieved in animals fed low protein-to-carbohydrate ratios with a negligible influence of total calorie intake.[44] The relationships between these dietary manipulations and the signaling pathways implicated in nutrient sensing and aging such as insulin and mechanistic target of rapamycin (mTOR) activity were assessed. Lower circulating insulin levels and inhibition of mTOR activity are associated with longevity.[45,46] While circulating insulin levels were influenced by both protein and carbohydrate concentrations, the lowest levels of insulin were measured when dietary protein was kept at the minimum. On the contrary, protein intake only had a small effect on mTOR activity. However, circulating glucose and branch-chain amino acids have strong effects on mTOR activation, and again, the lowest mTOR activity correlates with low protein-to-carbohydrate ratios. Finally, mitochondrial function is improved under low-protein diets.[44] Taken together, these results provide convincing evidence that enhanced longevity and metabolic fitness are attributable to low-protein diets in mice. These results also highlight the importance of balanced macronutrients beyond CR on energy intake, health, and longevity.

6. CIRCADIAN REGULATION AND CR

Much advance has been made in the last 2 decades in elucidating the mechanisms whereby CR exerts its positive effects on longevity and healthspan. While the precise molecular mechanisms are yet to be determined, three cellular pathways including the mTOR pathway, the insulin/IGF-1 system, and the sirtuin-regulated pathway have been shown to be conserved in mediating CR effects in model organisms.[47,48] More recently, emerging research has revealed a new and exciting connection between CR mediators and circadian rhythm.

The central and peripheral circadian clock systems coordinate an internal rhythm to maintain many aspects of physiology and behavior. Metabolic homeostasis and circadian clocks are interdependent and feeding is one of the stimuli capable of entraining peripheral clocks.[49] For example, time-restricted feeding of nocturnal mice during daytime shifted the phase of several peripheral clocks significantly within 1 week of feeding, including the liver clock, which exhibited a rapid phase shift within 2 days of altered feeding schedule.[50,51] To investigate a role of circadian regulation in CR-mediated longevity and health benefits, Katewa *et al.* demonstrated that in *Drosophila* mutations of the core circadian molecules *timeless* or *period* attenuated DR-mediated lifespan extension.[52] In addition, consistent with a link between circadian regulation and metabolism, mutations in *timeless* or *period* abolish the changes in fat turnover upon DR, whereas overexpression of *timeless* in peripheral tissues improves fat metabolism and extends lifespan of flies fed on control diet. As overexpression of *timeless* enhances the amplitude of its rhythmicity, these results suggest an intriguing notion that optimal circadian regulation may mimic DR in modulating metabolic control in promoting longevity.[52]

The mammalian circadian clock network consists of transcriptional feedback loops including transcription factors CLOCK and BMAL1, which form a complex and regulate the transcription of other clock genes, Periods (Per1 and Per2) and Cryptochromes (Cry1 and Cry2), whose activities, in turn, inhibit CLOCK:BMAL1 transcriptional activity forming a negative feedback mechanism. The expression of *Bmal1* can also be controlled by molecules constituting an "accessory loop."[53] Consistent with fly studies,[52] interactions between clock genes and CR effects in rodents have been demonstrated. First, CR significantly affects the rhythmicity of expression of circadian clock genes at both transcription and translation levels, resulting in modulated expression of several clock-controlled transcriptional factors and putative longevity genes.[54] Second, CR-dependent effects on some clock gene expression were absent in the liver of mice deficient for *Bmal1* suggesting both BMAL1-dependent and BMAL1-independent mechanisms in orchestration of the circadian clock network.[55] Third, the importance of BMAL1-dependent clock function in CR was illustrated in *Bmal1*-deficient mice, which failed to respond to CR-mediated lifespan extension.[55] While CR significantly reduces circulating IGF-1 and insulin levels in wild-type mice, such responses were substantially diminished in *Bmal1*$^{-/-}$ mice, suggesting BMAL1-dependent circadian clock function as an essential mediator of CR.

7. CR-MIMICKING INTERVENTIONS ON HEALTHSPAN

The health benefits of CR have been extensively tested and realized in a wide range of organisms.[39] The CALERIE study reported partial adherence to the intended 25% reduction in energy intake which ultimately hindered the full appreciation of CR-mediated health benefits in humans. Such practical limitation has necessitated the need for CR mimetics. To this end, mechanistic delineation of key pathways mediating the effects of CR has aided the identification of molecules such as sirtuin activators, including resveratrol[56] and SRT1720[57] as well as the mTOR inhibitor rapamycin[58] as potential surrogates for CR effects. To determine its lifespan effects in vivo, rapamycin was given to 20-month-old mice at three study sites where a 9 and 14% increase in lifespan was achieved in male and female animals, respectively.[58] While resveratrol improves cardiovascular and metabolic functions in mice,[59] no lifespan extension was observed unless the animals were under the metabolic stress such as high-fat diets.[60] On the contrary, supplementation with the more potent SIRT1-specific activator SRT1720 had extended lifespan and delayed onset of age-associated metabolic disorders in mice fed standard diet.[57] In a small, randomized double-blind crossover study, 11 healthy, obese men taking resveratrol for 30 days were found to have reduced resting metabolic rate, improved muscle mitochondrial respiration and additional metabolic changes mimicking CR effects.[61] Therefore, while the cellular mechanisms remain to be fully elucidated, boosting the activity of sirtuins through pharmacological means offers a great potential to positively impact health and longevity.

Nicotinamide adenine dinucleotide (NAD^+) is increasingly recognized as an important molecule that links cellular metabolism with regulation of aging. It was reported that the age-associated decline in NAD^+ levels negatively affects SIRT1 activity resulting in a pseudohypoxic state with decreased mitochondrial function and increased inflammation.[62] The potential of falling NAD^+ levels as a driver of aging (at least in some tissues) has been tested with promising outcomes. For example, feeding old mice the NAD^+ precursor, nicotinamide mononucleotide (NMN) not only restores NAD^+ to youthful levels but also reverses mitochondrial defects in old muscle.[62] A similar approach of long-term feeding of NMN has been shown to ameliorate age-associated physiological decline in mice.[63] Finally, in a neurodegenerative model, replenishment of intracellular NAD^+ reduces neuropathology, delays memory loss, and extends the lifespan of the animal.[64] Taken together, these results demonstrate the translational potential for manipulating NAD^+ levels to improve cellular function and healthspan in humans. To explain why NAD^+ levels decline during aging, it is hypothesized that since the critical NAD^+ synthesis enzyme nicotinamide phosphoribosyl transferase (NAMPT) is controlled by the circadian regulators CLOCK/BMAL1 and the amplitude of the clock network weakens during aging in mammals, a decline in NAMPT levels may account for decreased levels of NAD^+ with age.[65] Further studies are needed to understand the cause of age-associated decline in NAD^+ levels and whether increasing NAD^+ levels can be an antiaging intervention.

In conclusion, from model systems to initial human trials, several dietary interventions mimicking the effects of CR have shown great promise in simultaneously improving metabolic function and providing overall health benefits. However, much more basic research is needed to identify efficacious therapeutics that can be safely and practically translated to human studies in promoting healthspan.

REFERENCES

1. McCay CM, Crowell MF, Maynard LA. The effect of retarded growth upon the length of life span and upon the ultimate body size. *J Nutr.* 1935;10:63–79.
2. Fontana L, Partridge L. Promoting health and longevity through diet: from model organisms to humans. *Cell.* 2015;161(1):106–118.
3. Guarente L. Sir2 links chromatin silencing, metabolism, and aging. *Genes Dev.* 2000; 14(9):1021–1026.

4. Weindruch R, Walford RL. Dietary restriction in mice beginning at 1 year of age: effect on life-span and spontaneous cancer incidence. *Science.* 1982;215:1415–1418.
5. Clancy DJ, Gems D, Harshman LG, Oldham S, Stocker H, Hafen E, Leevers SJ, Partridge L. Extension of life-span by loss of CHICO, a *Drosophila* insulin receptor substrate protein. *Science.* 2001;292(5514):104–106.
6. Lee GD, Wilson MA, Zhu M, Wolkow CA, de Cabo R, Ingram DK, Zou S. Dietary deprivation extends lifespan in *Caenorhabditis elegans. Aging Cell.* 2006;5:515–524.
7. Kaeberlein TL, Smith ED, Tsuchiya M, Welton KL, Thomas JH, Fields S, Kennedy BK, Kaeberlein M. Lifespan extension in *Caenorhabditis elegans* by complete removal of food. *Aging Cell.* 2006;5:487–494.
8. Lin SJ, Kaeberlein M, Andalis AA, Sturtz LA, Defossez PA, Culotta VC, Fink GR, Guarente L. Calorie restriction extends *Saccharomyces cerevisiae* lifespan by increasing respiration. *Nature.* 2002;418(6895):344–348.
9. Houthoofd K, Vanfeteren JR. Longevity effect of dietary restriction in *Caenorhabditis elegans. Exp Gerontol.* 2006;41:1026–1032.
10. Piper MD, Blanc E, Leitão-Gonçalves R, Yang M, He X, Linford NJ, Hoddinott MP, Hopfen C, Soultoukis GA, Niemeyer C, Kerr F, Pletcher SD, Ribeiro C, Partridge L. A holidic medium for *Drosophila melanogaster. Nat Methods.* 2014;11(1):100–105.
11. Liao CY, Rikke BA, Johnson TE, Diaz V, Nelson JF. Genetic variation in the murine lifespan response to dietary restriction: from life extension to life shortening. *Aging Cell.* 2010;9(1):92–95.
12. Bishop NA, Guarente L. Two neurons mediate diet-restriction-induced longevity in *C. elegans. Nature.* 2007;447:545–549.
13. Magwere T, Goodall S, Skepper J, Mair W, Brand MD, Partridge L. The effect of dietary restriction on mitochondrial protein density and flight muscle mitochondrial morphology in *Drosophila. J Gerontol A Biol Sci Med Sci.* 2006;61:36–47.
14. Higami Y, Pugh TD, Page GP, Allison DB, Prolla TA, Weindruch R. Adipose tissue energy metabolism: altered gene expression profile of mice subjected to long-term caloric restriction. *FASEB J.* 2004;18:415–417.
15. Colman RJ, Anderson RM, Johnson SC, Kastman EK, Kosmatka KJ, Beasley TM, Allison DB, Cruzen C, Simmons HA, Kemnitz JW, Weindruch R. Caloric restriction delays disease onset and mortality in rhesus monkeys. *Science.* 2009;325(5937):201–204.
16. Colman RJ, Beasley TM, Kemnitz JW, Johnson SC, Weindruch R, Anderson RM. Caloric restriction reduces age-related and all-cause mortality in rhesus monkeys. *Nat Commun.* 2014;5:557.
17. Mattison JA, Roth GS, Beasley TM, Tilmont EM, Handy AM, Herbert RL, Longo DL, Allison DB, Young JE, Bryant M, Barnard D, Ward WF, Qi W, Ingram DK, de Cabo R. Impact of caloric restriction on health and survival in rhesus monkeys from the NIA study. *Nature.* 2012;489:318–321.
18. Mattison JA, Colman RJ, Beasley TM, Allison DB, Kemnitz JW, Roth GS, Ingram DK, Weindruch R, de Cabo R, Anderson RM. Caloric restriction improves health and survival of rhesus monkeys. *Nat Commun.* 2017;8:14063.
19. Redman LM, Heilbronn LK, Martin CK, Alfonso A, Smith SR, Ravussin E. Effect of calorie restriction with or without exercise on body composition and fat distribution. *J Clin Endocrinol Metab.* 2007;92:865–872.
20. Heilbronn LK, de Jonge L, Frisard MI, DeLany JP, Larson-Meyer E, Rood J, et al. Effect of 6-month calorie restriction on biomarkers of longevity, metabolic adaptation, and oxidative stress in overweight individuals: a randomized controlled trial. *JAMA.* 2006;295:1539–1548.
21. Das SK, Gilhooly CH, Golden JK, Pittas AG, Fuss PJ, Cheatham RA, et al. Long-term effects of 2 energy-restricted diets differing in glycemic load on dietary adherence, body

composition, and metabolism in CALERIE: a 1-y randomized controlled trial. *Am J Clin Nutr.* 2007;84:1023–1030.
22. Racette SB, Weiss EP, Villareal DT, Arif H, Steger-May K, Schechtman KB, et al. One year of caloric restriction in humans: feasibility and effects on body composition and abdominal adipose tissue. *J Gerontol A Biol Sci.* 2006;61:943–950.
23. Rochon J, Bales CW, Ravussin E, Redman LM, Holloszy JO, Racette SB, Roberts SB, Das SK, Romashkan S, Galan KM, Hadley EC, Kraus WE. CALERIE Study Group. Design and conduct of the CALERIE study: comprehensive assessment of the long-term effects of reducing intake of energy. *J Gerontol A Biol Sci Med Sci.* 2011;66(1):97–108.
24. Redman LM, Kraus WE, Bhapkar M, Das SK, Racette SB, Martin CK, Fontana L, Wong WW, Roberts SB, Ravussin E, et al. Energy requirements in nonobese men and women: results from CALERIE. *Am J Clin Nutr.* 2014;99:71–78.
25. Pieper C, Redman L, Racette S, Roberts S, Bhapkar M, Rochon J, Martin C, Kraus W, Das S, Williamson D, et al. Development of adherence metrics for caloric restriction interventions. *Clin Trials.* 2011;8:155–164.
26. Wong WW, Robert SB, Racette SB, Das SK, Redman LM, Rochon J, Bhapkar MV, Clarke LL, Kraus WE. The doubly labeled water method produces highly reproducible longitudinal results in nutrition studies. *J Nutr.* 2014;144:777–783.
27. Racette SB, Das SK, Bhapkar M, Hadley EC, Roberts SB, Ravussin E, Pieper C, DeLany JP, Kraus WE, Rochon J, et al. Approaches for quantifying energy intake and %calorie restriction during calorie restriction interventions in humans: the multicenter CALERIE study. *Am J Physiol Endocrinol Metab.* 2012;302:E441–E448.
28. Ravussin E, Redman LM, Rochon J, Das SK, Fontana L, Kraus WE, Romashkan S, Williamson DA, Meydani SN, Villareal DT, et al. A 2-year randomized controlled trial of human caloric restriction: feasibility and effects on predictors of health span and longevity. *J Gerontol A Biol Sci Med Sci.* 2015;70:1097–1104.
29. Das SK, Roberts SB, Bhapkar MV, Villareal DT, Fontana L, Martin CK, Racette SB, Fuss PJ, Kraus WE, Wong WW, Saltzman E, Pieper CF, Fielding RA, Schwartz AV, Ravussin E, Redman LM. CALERIE-2 Study Group.Body-composition changes in the Comprehensive Assessment of Long-term Effects of Reducing Intake of Energy (CALERIE)-2 study: a 2-y randomized controlled trial of calorie restriction in nonobese humans. *Am J Clin Nutr.* 2017;105(4):913–927.
30. Longo VD, Mattson MP. Fasting: molecular mechanisms and clinical applications. *Cell Metab.* 2014;19(2):181–192.
31. Gonidakis S, Finkel SE, Longo VD. Genome-wide screen identifies Escherichia coli TCA-cycle-related mutants with extended chronological lifespan dependent on acetate metabolism and the hypoxia-inducible transcription factor ArcA. *Aging Cell.* 2010;9 (5):868–881.
32. Longo VD, Ellerby LM, Bredesen DE, Valentine JS, Gralla EB. Human Bcl-2 reverses survival defects in yeast lacking superoxide dismutase and delays death of wild-type yeast. *J Cell Biol.* 1997;137(7):1581–1588.
33. Kaeberlein TL, Smith ED, Tsuchiya M, Welton KL, Thomas JH, Fields S, Kennedy BK, Kaeberlein M. Lifespan extension in *Caenorhabditis elegans* by complete removal of food. *Aging Cell.* 2006;5(6):487–494.
34. Goodrick CL, Ingram DK, Reynolds MA, Freeman JR, Cider N. Effects of intermittent feeding upon body weight and lifespan in inbred mice: interaction of genotype and age. *Mech Ageing Dev.* 1990;55(1):69–87.
35. Mattson MP, Longo VD, Harvie M. Impact of intermittent fasting on health and disease processes. *Ageing Res Rev.* 2017;39:46–58.
36. Anson RM, Guo Z, de Cabo R, Iyun T, Rios M, Hagepanos A, Ingram DK, Lane MA, Mattson MP. Intermittent fasting dissociates beneficial effects of dietary restriction on

glucose metabolism and neuronal resistance to injury from calorie intake. *Proc Natl Acad Sci USA*. 2003;100(10):6216–6220.
37. Mager DE, Wan R, Brown M, Cheng A, Wareski P, Abernethy DR, Mattson MP. Caloric restriction and intermittent fasting alter spectral measures of heart rate and blood pressure variability in rats. *FASEB J*. 2006;20(6):631–637.
38. Wan R, Ahmet I, Brown M, Cheng A, Kamimura N, Talan M, Mattson MP. Cardioprotective effect of intermittent fasting is associated with an elevation of adiponectin levels in rats. *J Nutr Biochem*. 2010;21(5):413–417.
39. Fontana L, Partridge L, Longo VD. Extending healthy life span—from yeast to humans. *Science*. 2010;328(5976):321–326.
40. Thissen JP, Pucilowska JB, Underwood LE. Differential regulation of insulin-like growth factor I (IGF-I) and IGF binding protein-1 messenger ribonucleic acids by amino acid availability and growth hormone in rat hepatocyte primary culture. *Endocrinology*. 1994;134(3):1570–1576.
41. Brandhorst S, Choi IY, Wei M, Cheng CW, Sedrakyan S, Navarrete G, Dubeau L, Yap LP, Park R, Vinciguerra M, Di Biase S, Mirzaei H, Mirisola MG, Childress P, Ji L, Groshen S, Penna F, Odetti P, Perin L, Conti PS, Ikeno Y, Kennedy BK, Cohen P, Morgan TE, Dorff TB, Longo VD. A periodic diet that mimics fasting promotes multi-system regeneration, enhanced cognitive performance, and healthspan. *Cell Metab*. 2015;22(1):86–99.
42. Mair W, Piper MD, Partridge L. Calories do not explain extension of life span by dietary restriction in *Drosophila*. *PLoS Biol*. 2005;3:e223.
43. Lee KP, Simpson SJ, Clissol FJ, Broo R, Ballard JW, Taylor PW, Soran N, Raubenheimer D. Lifespan and reproduction in *Drosophila*: new insights from nutritional geometry. *Proc Natl Acad Sci USA*. 2008;105:2498–2503.
44. Solon-Biet SM, McMahon AC, Ballard JW, Ruohonen K, Wu LE, Cogger VC, Warren A, Huang X, Pichaud N, Melvin RG, Gokarn R, Khalil M, Turner N, Cooney GJ, Sinclair DA, Raubenheimer D, Le Couteur DG, Simpson SJ. The ratio of macronutrients, not caloric intake, dictates cardiometabolic health, aging, and longevity in ad libitum-fed mice. *Cell Metab*. 2014;19:418–430.
45. Kapahi P, Chen D, Rogers AN, Katewa SD, Li PW, Thomas EL, Kockel L. With TOR, less is more: a key role for the conserved nutrient-sensing TOR pathway in aging. *Cell Metab*. 2010;11:453–465.
46. Gallinetti J, Harputlugil E, Mitchell JR. Amino acid sensing in dietary-restriction-mediated longevity: roles of signal-transducing kinases GCN2 and TOR. *Biochem J*. 2013;449:1–10.
47. Finkel T. The metabolic regulation of aging. *Nat Med*. 2015;21(12):1416–1423.
48. Mercken EM, Hu J, Krzysik-Walker S, Wei M, Li Y, McBurney MW, de Cabo R, Longo VD. SIRT1 but not its increased expression is essential for lifespan extension in caloric-restricted mice. *Aging Cell*. 2014;13:193–196.
49. Green CB, Takahashi JS, Bass J. The meter of metabolism. *Cell*. 2008;728–742.
50. Damiola F, Le Minh N, Preitner N, Kornmann B, Fleury-Olela F, Schibler U. Restricted feeding uncouples circadian oscillators in peripheral tissues from the central pacemaker in the suprachiasmatic nucleus. *Genes Dev*. 2000;14:2950–2961.
51. Stokkan KA, Yamazaki S, Tei H, Sakaki Y, Menaker M. Entrainment of the circadian clock in the liver by feeding. *Science*. 2001;291(5503):490–493.
52. Katewa SD, Akagi K, Bose N, Rakshit K, Camarella T, Zheng X, Hall D, Davis S, Nelson CS, Brem RB, Ramanathan A, Sehgal A, Giebultowicz JM, Kapahi P. Peripheral circadian clocks mediate dietary restriction-dependent changes in lifespan and fat metabolism in *Drosophila*. *Cell Metab*. 2015;23:143–154.

53. Sato TK, Yamada RG, Baggs JE, Miraglia LJ, Kobayashi TJ, Welsh DK, Kay SA, Ueda HR, Hogenesch JB. Feedback repression is required for mammalian circadian clock function. *Nat Genet.* 2006;38(3):312–319.
54. Patel SA, Velingkaar N, Makwana K, Chaudhari A, Kondratov R. Calorie restriction regulates circadian clock gene expression through BMAL1 dependent and independent mechanisms. *Sci Rep.* 2016;6:25970.
55. Patel SA, Chaudhari A, Gupta R, Velingkaar N, Kondratov RV. Circadian clocks govern calorie restriction-mediated life span extension through BMAL1- and IGF-1-dependent mechanisms. *FASEB J.* 2016;30:1634–1642.
56. Howitz KT, Bitterman KJ, Cohen HY, Lamming DW, Lavu S, Wood JG, Zipkin RE, Chung P, Kisielewski A, Zhang LL, Scherer B, Sinclair DA. Small molecule activators of sirtuins extend *Saccharomyces cerevisiae* lifespan. *Nature.* 2003;425:191–196.
57. Mitchell SJ, Martin-Montalvo A, Mercken EM, Palacios HH, Ward TM, Abulwerdi G, Minor RK, Vlasuk GP, Ellis JL, Sinclair DA, Dawson J, Allison DB, Zhang Y, Becker KG, Bernier M, de Cabo R. The SIRT1 activator SRT1720 extends lifespan and improves health of mice fed a standard diet. *Cell Rep.* 2014;6:836–843.
58. Harrison DE, Strong R, Sharp ZD, Nelson JF, Astle CM, Flurkey K, Nadon NL, Wilkinson JE, Frenkel K, Carter CS, Pahor M, Javors MA, Fernandez E, Miller RA. Rapamycin fed late in life extends lifespan in genetically heterogeneous mice. *Nature.* 2009;460:392–395.
59. Lagouge M, Argmann C, Gerhart-Hines Z, Meziane H, Lerin C, Daussin F, Messadeq N, Milne J, Lambert P, Elliott P, Geny B, Laakso M, Puigserver P, Auwerx J. Resveratrol improves mitochondrial function and protects against metabolic disease by activating SIRT1 and PGC-1α. *Cell.* 2006;127:1109–1122.
60. Pearson KJ, Baur JA, Lewis KN, Peshkin L, Price NL, Labinskyy N, Swindell WR, Kamara D, Minor RK, Perez E, Jamieson HA, Zhang Y, Dunn SR, Sharma K, Pleshko N, Woollett LA, Csiszar A, Ikeno Y, Le Couteur D, Elliott PJ, Becker KG, Navas P, Ingram DK, Wolf NS, Ungvari Z, Sinclair DA, de Cabo R. Resveratrol delays age-related deterioration and mimics transcriptional aspects of dietary restriction without extending life span. *Cell Metab.* 2008;8:157–168.
61. Timmers S, Konings E, Bilet L, Houtkooper RH, van de Weijer T, Goossens GH, Hoeks J, van der Krieken S, Ryu D, Kersten S, Moonen-Kornips E, Hesselink MK, Kunz I, Schrauwen-Hinderling VB, Blaak EE, Auwerx J, Schrauwen P. Calorie restriction-like effects of 30 days of resveratrol supplementation on energy metabolism and metabolic profile in obese humans. *Cell Metab.* 2011;14:612–622.
62. Gomes AP, Price NL, Ling AJ, Moslehi JJ, Montgomery MK, Rajman L, White JP, Teodoro JS, Wrann CD, Hubbard BP, Mercken EM, Palmeira CM, de Cabo R, Rolo AP, Turner N, Bell EL, Sinclair DA. Declining NAD+ induces a pseudohypoxic state disrupting nuclear-mitochondrial communication during aging. *Cell.* 2013;155:1624–1638.
63. Mills KF, Yoshida S, Stein LR, Grozio A, Kubota S, Sasaki Y, Redpath P, Migaud ME, Apte RS, Uchida K, Yoshino J, Imai SI. Long-term administration of nicotinamide mononucleotide mitigates age-associated physiological decline in mice. *Cell Metab.* 2016;24:795–806.
64. Fang EF, Kassahun H, Croteau DL, Scheibye-Knudsen M, Marosi K, Lu H, Shamanna RA, Kalyanasundaram S, Bollineni RC, Wilson MA, Iser WB, Wollman BN, Morevati M, Li J, Kerr JS, Lu Q, Waltz TB, Tian J, Sinclair DA, Mattson MP, Nilsen H, Bohr VA. NAD+ replenishment improves lifespan and healthspan in ataxia telangiectasia models via mitophagy and DNA repair. *Cell Metab.* 2016;24:566–581.
65. Guarente L. Aging research—where do we stand and where are we going? *Cell.* 2014;159:15–19.

CHAPTER THREE

Obesity, Metabolism, and Aging: A Multiscalar Approach

R. Alexander Bentley*,[1], Corinna N. Ross†, Michael. J. O'Brien†
* University of Tennessee, Knoxville, TN, United States
† Texas A&M University–San Antonio, San Antonio, TX, United States
[1] Corresponding author. E-mail address: rabentley@utk.edu

Contents

Abstract

Obesity contributes to the aging process through the alteration of metabolic pathways evidenced biochemically in the relationship between caloric restriction and longevity. Humans have entered into an era of metabolism and aging entirely unprecedented in their evolution, with a diet that, for many, contains a majority of calories as sugar and yields an expected lifespan of over 80 years in industrialized nations. Deeply embedded in the complex issue of obesity are questions of behavior, causality versus correlation, and appropriate models. For example, are primates a better reference than mice for studying metabolic connections between obesity and aging? We consider those issues from the standpoint of life-history theory, especially implications of the interplay of refined sugar and socioeconomic disparities for the future of human health.

1. INTRODUCTION: OBESITY, AGING, AND METABOLISM

Obesity is a significant public-health concern[1,2] with two-thirds of Americans classified as overweight and over 100,000 US deaths per year attributed to it. Obesity and metabolic syndrome are considered as

Progress in Molecular Biology and Translational Science, Volume 155
ISSN 1877-1173
http://dx.doi.org/10.1016/bs.pmbts.2017.11.016

contributors to early mortality, especially from type 2 diabetes and cardiovascular disease. In addition, the metabolic dysfunction and cellular-pathway functional cascades are associated with longevity as evidenced biochemically by the finding that caloric restriction increases life span in animal-model systems.[3]

The relationship between obesity and aging is a multiscalar problem, which for heuristic purposes we represent as an hourglass (Fig. 1). In terms of temporal scale, the hourglass could simultaneously remind us that the metabolic basis for obesity and aging depends on processes that range from molecular reactions to short-term human decisions, long-term habits, policies, and traditions among populations that ultimately are transformed on an evolutionary timescale. In terms of physical scale, the hourglass represents the

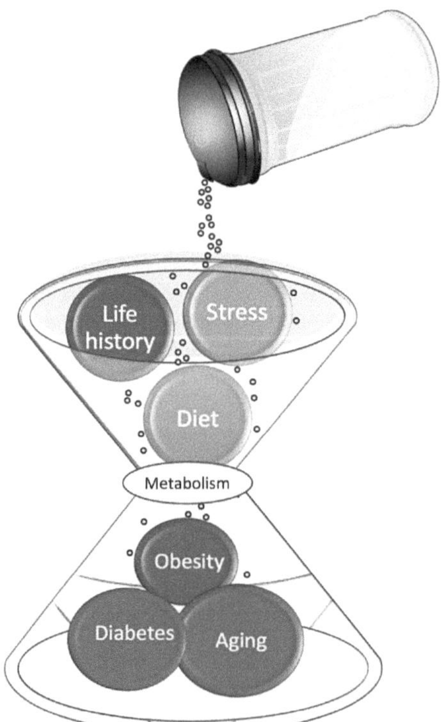

Fig. 1 The multiscalar relationship between obesity and aging, represented as an hourglass. Metabolism occupies the microscale interface between the societal inputs at one macro-scale and phenotypic outcome at the other macro-scale. In the past 50 years, refined sugar has become a major contributor to altering this process.

range from biomolecular interactions at the narrowest point, to individuals, and then to governments at the widest points, where obesity and diabetes pose a global problem in developed nations.[4] Commercial-food industries, particularly those involving refined sugar, feed (literally) into dietary decisions within populations.

At the vertex of the hourglass is metabolism, which is the biomolecular reactions that translate these societal-level forces and individual dietary choices into phenotypic outcomes that include obesity, diabetes, and aging. Obesity alters the metabolism of glucose, fatty acids, and amino acids, which leads to a decrease in insulin sensitivity and hence a decline in the ability to adjust to energy availability.[5] There is reason to believe that obesity contributes directly to the aging process, as the metabolic effects of aging include altered mitochondrial function and degraded function of nutrient-signaling pathways that regulate the balance of insulin versus glucagon in maintaining blood glucose.[6] These compounds for nutrient signaling include protein complexes for the mechanistic target of rapamycin (mTOR), which senses energy levels within the cell and subsequently moderates metabolism, as well as sirtuins, which are enzymatic proteins involved in the metabolic sensor 5′ adenosine monophosphate-activated protein kinase (AMPK), which can increase cellular energy supply and decrease energy demand.[7–10]

These nutrient-sensing and longevity pathways constitute an interactive caloric-reduction sensing network that alters the involvement of sirtuins, mTOR, and AMPK in the regulation of autophagy,[10] the cell mechanism that disassembles and recycles cellular components and thus has a role in lifespan extension.[11] Autophagy is positively regulated by the sirtuins and negatively by mTOR. The latter regulates protein synthesis and controls the secretion of senescence-associated secretory phenotype (SASP), which makes neighboring cells senescent and degrades the tissue microenvironment.[9] The three pathways may comprise a caloric-restriction-sensing network, in that both mTOR and AMPK can regulate a kinase (ULK1) that aids in forming the double membrane that marks an organelle for autophagic destruction.[10] AMPK is activated when energy supply is low, which slows aging and increases resistance to pathologies of old age.[7]

In mice, caloric restriction can extend lifespan by 50%.[12] The extended lifespan in mice that have had their growth-hormone receptor gene disrupted appears to result from increased oxidative metabolism combined with enhanced fatty-acid oxidation.[13] Most of the weight loss from caloric restriction tends to come from visceral fat stores.[10] The best longevity gains of caloric reduction occur in mice that also maintain fat stores despite the 40%

caloric reduction.[14] Nevertheless, people and mice are separated by about hundred million years of evolution, with key differences in responses to caloric restriction. Calorically restricted mice, for example, have lower body temperatures whereas humans under 25% caloric restriction do not.[15] Pertinent differences between humans and laboratory rodents include gut morphology, metabolic physiology, feeding behaviors, stress reactivity, and reproductive timing and investment.[16–18] Primates, on the other hand, share a closer evolutionary history with humans, offering more insight for translational biomedical research.[19] Progression of disease associated with both obesity and aging have been evaluated in nonhuman primates, such as macaques.[20,21] Diet-induced obesity is associated with metabolic dysfunction, increased risk of cardiovascular disease, and increased mortality in primates.[22–24] As they age, primates display cognitive decline, increased cardiovascular disease, and kidney disease.[25–27] Among macaques, long-term caloric restriction intervention is associated with decreased prevalence of cardiovascular disease, type 2 diabetes, and neoplasias.[28,29]

Due to their short lifespan and fast reproduction,[30] marmosets, which are small New World monkeys, are excellent models for studies of how the functional decline of aging relates to obesity. There is a long history of studies of marmoset hormonal development, reproduction, cooperative breeding behavior, and use as a biomedical model. This has resulted in a large base of average values for infant growth, body weight, and hematological measures.[30] At the molecular scale, obese adult marmosets are more likely to suffer from metabolic dysfunction, including increased insulin resistance, increased serum-triglyceride concentrations, increased fasting glucose concentrations, and, in rare cases, type 2 diabetes.[17] Phenotypic changes associated with aging in the marmoset include decreased serum albumin concentrations, increased β-amyloid deposition in the brain, decreased neurogenesis, and fat mass, degeneration of bone structure, and increased prevalence of neoplasias.[30–34] Metabolite-system connectivity significantly decreases in aging marmosets, and changes in the production of specific metabolites[35] may identify pathways of change for targeted intervention. In marmoset colonies, mortality resulting from cardiovascular events and amyloidosis increases as obesity prevalence increases.[34] The interaction between metabolic dysfunction and obesity and the impact on marmoset health and aging remains unclear, but the establishment of this model organism and descriptions of typical phenotypic changes will allow us to bridge molecular, individual, and social scales to evaluate these factors in human aging.

2. SCALING OUT: LIFE-HISTORY THEORY

Caloric restriction is the standard recommendation for treating obesity, as it also improves insulin sensitivity.[5,36] Caloric restriction effectively resets an individual's energy balance from one that prioritizes biosynthesis and reproduction to one devoted mainly to maintenance, repair, and daily activities.[10] There is a similar trade-off in the longer evolutionary view. Moving out to the scale of ecology, this energy-partitioning theory is consistent with allometric theory for observed scaling laws, by which metabolic rate is proportional to body mass raised to an exponent of about three-fourths for placental mammals.[37–39] Although the effect of caloric restriction on basal metabolic rate (BMR) is complex, some studies do show reduced BMR with caloric reduction, and others show that caloric restriction decreases reliance on carbohydrates in favor of oxidizing fatty acids.[12]

Related models of allometric-scaling relationships exist for a wide range of organisms, based on the fundamental biophysics of how metabolic energy is allocated between maintenance of existing tissue and the production of new biomass.[40] Results suggest a general definition of biological time that is approximately invariant and common to all mammals,[41] such that a mammal's lifespan equates to about 1.5 billion beats of its heart. In other words, a slower heart rate means a longer lifespan in calendar time.

Controlling for body size, lifespan also correlates positively with brain size in mammals, including primates.[42] This scaling relationship has been explained by life-history theory (LHT), which is an analytical framework that examines how organisms should allocate time and energy to fitness maximization in the face of trade-offs.[43] Under LHT, the increasing risk of mortality during aging is seen as the result of genes that evolved for beneficial effects early in life, but deleterious effects later in life.[44] In other words, aging, or senescence, is the inevitable outcome of optimal allocation of time and energy over the life course to maximize reproduction. LHT addresses the evolutionary forces that shape the course of the lifelong energy budget involved in development, growth, reproduction, maintenance, and aging.[45,46]

LHT predicts that as external threats to mortality increase, aging accelerates to avoid "wasting" an energy budget allocated to the future.[43] There is a trade-off between quality and quantity of offspring, which means that survival to longer lifespans correlates with a decrease in number of offspring.[47] This might explain why, across societies, greater wealth correlates with fewer children whereas within societies more wealth means more children.[48]

3. OUR HUNTER–GATHERER PAST

LHT theorists posit that humans, unlike other primates, improve in their abilities to acquire and provision food (e.g., hunting, gathering, cooking) as they age. In traditional hunter–gatherer societies, humans produce less than they consume for about 20 years, and human net food production does not peak until age 45.[49] By age 15, chimpanzees on average have consumed 43% of their expected lifetime calories and produced 40%. In contrast, humans have consumed 22% and produced only 4%.[50] Under LHT, caloric restriction can delay menarche because the slowed growth and fat deposition continue longer to outweigh the cost of waiting to reproduce, despite the increased mortality risk of the poor.[43]

The prehistory of human diet is a story of the feedback among dietary changes that created evolutionary niches. Each new niche favored new means of obtaining and processing food and genetic changes for phenotypes that benefited nutritionally from them. Human physiology evolved around a diet containing very little sugar and virtually no refined carbohydrates. Hundreds of thousands of years ago, when cooking allowed the digestion of nutritious starchy tubers, the freeing up of energetic requirements for digestion allowed larger brains to evolve. Larger brains facilitated group cooperation and more-complex means of obtaining food, such as hunting in groups and strategic foraging.

Hunter–gatherers generally do not exhibit obesity, diabetes, or cardiovascular disease.[51,52] For example, Tsimane hunter–gatherer–fishers (and part-time farmers) of Bolivia exhibit few cardiovascular risk factors and virtually no obesity, hypertension, or high blood sugar, despite their highly infectious inflammatory burden.[52] In a study of over 700 Tsimane adults over 40 years old, only 3% showed signs of significant atherosclerotic disease in their coronary-artery calcium levels. The rate among those over 75 years old was lower than that for industrialized populations by a factor of five.[52]

As hunter–gatherer rhythms are more intermittent than modern diets, a reasonable hypothesis is that the frequency and circadian timing of energy intake affect aging and metabolism.[53] The idea is that intermittent energy restriction, such as fasting 2 days a week, induces mild stress that protects against more-severe stress. In contrast, fasting for just two-thirds of a day appears to engage fat metabolism (release of fatty acids from adipose tissues into the blood), and ketone production, with liver glycogen stores being depleted after about half a day.[53] Fasting also inhibits the nutrient-responsive mTOR pathway, which in turn stimulates the cellular-level process of removing damaged molecules and organelles in tissues, including the liver, kidney, and muscles.[53] Alternate-day fasting among adults, for example, yields similar weight reduction as caloric restriction and similar changes in lipid levels and insulin sensitivity, but fasting yields significantly larger decrease in fat mass and corresponding increase in lean mass than caloric restriction.[54] Fasting also stimulates the production of proteins that encourage mitochondrial functions, increasing insulin sensitivity and ketone levels, mobilizing fatty acids, and reducing overall levels of insulin, leptin, and proinflammatory cytokines.[53,55,56]

Constant energy intake may be incompatible with the evolved physiology of energy storage, including liver glycogen stores and the fatty acids in adipose tissue.[53] With three meals a day plus snacks, a steady supply of nutrients keeps cells synthesizing protein while suppressing autophagy.[12,53] This ultimately is a result of changes in the last 10,000 years, in the conversion to intensive, carbohydrate-rich diets first produced by Neolithic farmers. Carbohydrate content changed more than any other dietary component during the Neolithic transition. In Europe about 8000 years ago, human diets changed over just a few centuries from intermittent low-calorie meat, fruits, nuts, and tubers to the high-calorie foods of processed grains.

For Neolithic populations of Europe, West Africa, and the Middle East, significant supplements to these carbohydrate diets were dairy products, especially milk, which is a "nutritional bonanza" of fat, proteins, carbohydrates, vitamins, calcium, and even water, but only if it can be digested.[57] Milk contains the disaccharide sugar lactose, which is the primary osmotic regulator of milk synthesis and as such is typically the least variable macrocomponent of milk. For complete digestion, lactose must be broken down into the two monosaccharides from which it was synthesized, glucose and galactose. Most babies naturally produce the enzyme lactase so that they can take advantage of the nutrients in mother's milk, but for most of the world's population, lactase production shuts down in the postweaning years. Drinking milk after that leads to a battery of symptoms, including diarrhea, cramping, gas, nausea, and vomiting. For most modern populations, with access to medical care if needed, this is not much of a problem, but for those without access, untreated diarrhea, especially in juveniles, can be fatal.

In Europe, over many centuries populations evolved a single nucleotide polymorphism (−13910*T) located 13.9 kb upstream of the lactase (LCT) gene,[58] which allowed them to fully digest milk and other dairy products, which explains most of the modern distribution of lactase persistence (LP).[58] Whether the LP allele achieved high frequency depended on the probability of the offspring of milk drinkers becoming milk drinkers themselves.[59] In other words, milk drinking had to be not just a learned tradition but a reliably learned tradition.[60] If that probability was high, then a significant fitness advantage to LP individuals resulted in a high frequency of LP within a few hundred generations. Conversely, if a significant proportion of the offspring of milk drinkers did not drink milk themselves, then unrealistically strong selection for LP was needed for the allele to spread. The LP allele is one of the strongest cases of evolutionary selection for any gene in the human genome.[61,62] The nutritional benefits of dairy imposed a positive feedback loop between the selective advantages of lactose tolerance for humans consuming dairy products and the selective breeding of high milk yield in dairy cattle.[63]

Anthropologists once viewed the Neolithic as the driver of better ways of life for those who practiced agriculture, in that it led to the rise of villages and eventually cities and all their trappings, such as monumental architecture, but they overlooked the downside of civilization. Yes, the Neolithic supported larger populations through more-intensive land use,[64] but it also represented a step-down in terms of health, with a decrease in human stature, an increase in disease load on populations, and poorer nutrition than the balanced hunter–-gather diets. Human niche-constructing behaviors that have what seemingly are positive effects can, under changes in the niche, also have negative effects.[63] In most Western countries, for example, diet-related chronic diseases represent the largest single cause of morbidity and mortality.[65]

Although we often attempt to identify a single dietary factor as leading to a disease, it is becoming increasingly clear that almost all so-called diseases of civilization have multiple and complex dietary and genetic causes. Heart disease, for example, does not arise simply from excessive saturated fat in the diet but rather from a complex interaction of multiple foods that are part of our ecological inheritance from our Neolithic ancestors—dairy products, cereals, sugars, fatty meats, and salt. These foods, in turn, adversely influence proximate nutritional factors—glycemic load, fatty-acid composition, acid-base balance, sodium–potassium ratio, and fiber content—all of which underlie or exacerbate virtually the entire spectrum of chronic diseases of civilization.

Dietary changes since the agricultural revolution have likely triggered extensive selection on human genes expressed in metabolism, but such

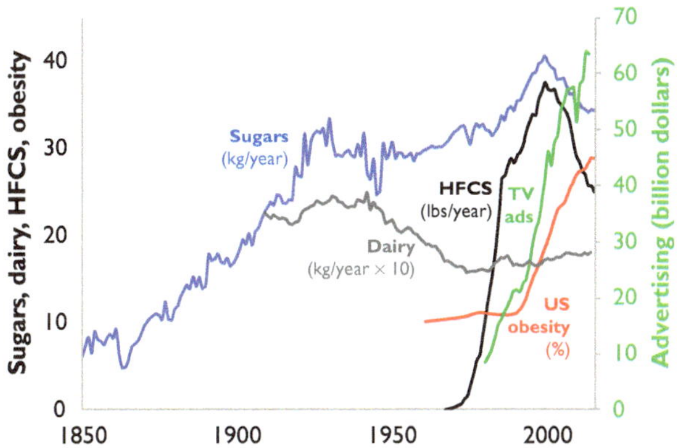

Fig. 2 A timeline of the increase in contribution of refined sugar and high fructose corn syrup (HFCS) to US diet, together with the increase in US obesity rate. The data for sugar, dairy and HFCS consumption per capita are from USDA Economic Research Service (https://www.usda.gov/topics/data) except for sugar consumption before 1967, which are historical estimates.[66] Obesity data (% of US adult population) are from the Robert Wood Johnson Foundation's Trust for America's Health (www.stateofobesity.org). Total US television advertising data are from the World Advertising Research Center (www.warc.com). The *y*-axis on the left covers all data series except advertising expenditures, which uses the *y*-axis on the right.

selection is likely to be ongoing and only partially compensatory for the ill effects of the diseases concerned. The ultimate factor underlying diseases of civilization remains at the intersection between our ancient genome with the nutritional qualities of recently introduced foods.[65] In other words, the agricultural niche developed so recently in terms of evolutionary time, not to mention its exponential growth rate and almost limitless cultural changes that it created, that even allowing for rapid genetic change, the human genome cannot keep up.[63] On an evolutionary time scale, these historical developments in human diet (Fig. 2) were like a gunshot. None was more significant than the one fired by sugar, especially if we consider the advent of high-fructose corn syrup in the past 50 years.

4. THE SUGAR EXPLOSION

Given the prehistory of how a two-molecule sugar changed human populations over millennia, consider how one-molecule sugars have changed much of the world's population over the past few centuries.

Archaeological evidence indicates that sugarcane evolved in South East Asia, about 8000 years ago, perhaps used to fatten pigs, but the earliest sign of sugar being chemically refined is in India about 2500 years ago.[67] From there, refined sugar spread across Eurasia, from China to Persia and eventually the Mediterranean by the 13th century, when it was still a rare and expensive spice rather than a common food.[67] By the late 15th century, large-scale cultivation and refinement of sugarcane for trade had begun on the Atlantic island of Madeira, and by the 17th century the Portuguese had instilled a slave-based plantation economy, leading to the Caribbean sugar monoculture that traded sugar to Western Europe during the 18th and 19th centuries.[68] The labor needed to cultivate the vast sugarcane plantations was found through a transatlantic slave trade involving at least 12 million people from Africa.[69] In terms of human health, one hypothesis is that markets of refined sugar actively drive the rapid growth in type 2 diabetes and coronary heart disease in countries around the world. As each nation undergoes a rise in sugar consumption, there is a corresponding rise in obesity rate. Granger-causality testing on cross-sectional data from 175 countries has shown that each increase of 150 kcal/person/day in sugar availability causes a 1% increase in national diabetes prevalence.[4] This hypothesis is supported by evidence that dietary sugar, especially as supplied by beverages, is both addictive[70] and a causal factor in high obesity rates in the developed world.[4,71,72] Westerners now get about 15% of their total calories from sugar, and sugar (cane and beets) is the world's seventh-ranked crop by cultivated land area. The widespread marketing of inexpensive high-sugar foods, especially sugar-sweetened beverages, may drive the rise in obesity,[73] diabetes,[4,74] nonalcoholic fatty liver disease,[75] and heart disease.[76] Even laboratory rodents and household pets are becoming obese.[77,78]

In evolutionary psychology, the follow-on from LHT is the hypothesis that fat serves as an "insurance" against food scarcity or food insecurity[79,80] and that modern obesity is a result of high availability of food compared to the conditions of ancient societies and human evolutionary origins.[81] Under this view, obesity is driven by the commercial oversupply and widespread marketing of inexpensive high-sugar foods, especially sugar-sweetened beverages, that may drive the rise in obesity,[73,82,83] diabetes,[4] and heart disease.[76] Fructose, which has steadily replaced other forms of sugar in foods, especially beverages, since the early 1970s, may have played a substantial role in the rise of obesity and diabetes.[83–86] Nevertheless, the underlying causes of the rapid increase in US obesity since the 1980s, which includes childhood obesity, are complex.[87] In 1990, about 11% of a typical US state population was obese,

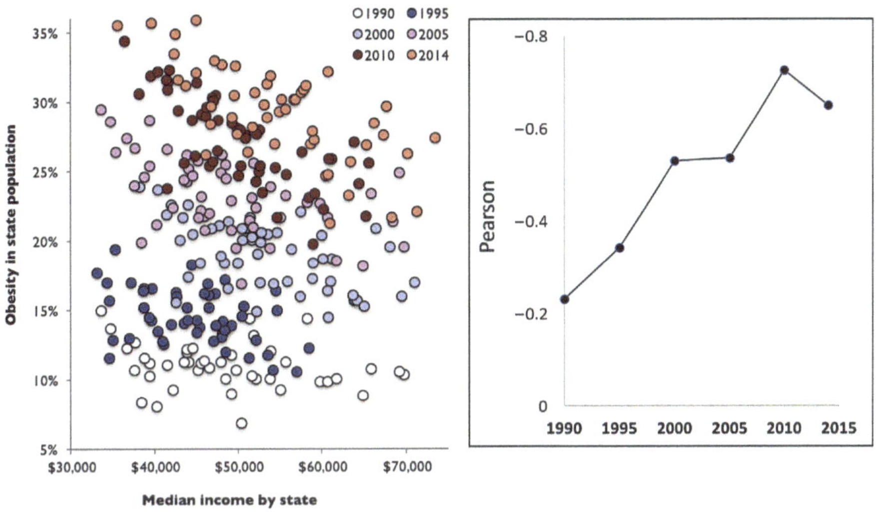

Fig. 3 The inverse correlation between median household income and obesity in US states, for different decades since 1990. Each datum point represents a statewide average for both statistics for the year. Median-income data are from the National Center for Education Statistics (www.nces.ed.gov), and obesity data are from the Robert Wood Johnson Foundation's Trust for America's Health (www.stateofobesity.org).

and no state had higher than 15% obesity in its adult population (Fig. 3). By 2014, statewide obesity rates had nearly tripled, and *no* state was below 20% obese. To put this in perspective, any state in 2015 would have registered as an extreme outlier in 1990. The alarming rise in childhood obesity suggests changes in activity, dietary intake, and parental perceptions of obesity.[88,89] In developed economies over the past half century, abundant calories, combined with sedentary lifestyles and automobiles, have brought about the correlated epidemics of obesity and diabetes.[53] Although social and cultural influence are definitely among the risk factors for obesity,[90–92] debate continues over whether this "social influence" is merely the tendency for like-minded individuals to coassociate.[93]

Studies of longevity often record the lifespans of family members,[94] so that we can get some idea of how both genes and behaviors are inherited. Dietary habits, including sugar consumption, are developed early in childhood and become part of family routines.[95] Higher obesity rates correlate with exposure to childhood poverty, with variation by ethnicity and gender.[96] Weaning appears to be a crucial time for the propensity for childhood obesity. In infant marmosets, observations of behavior and solid and liquid feeding show that the ones who become obese (>14% body fat) by 1-year of age were the ones who had been weaned earlier, such that the obese infants began eating solid food sooner and drinking more grams of diet in each liquid feeding episode.[89]

5. THE SOCIONOMICS OF AGING AND METABOLISM

As "diseases of affluence" accompany economic growth and increased availability of sugar-sweetened beverages, obesity and diabetes increase accordingly,[4] often disproportionately affecting the poorest classes. Modern obesity, which disproportionately affects the poor, might result from the "double-whammy" of cheap calories in food deserts available to those who are under the psychosocial stresses of poverty, such as family instability.[79,96–98] As poverty impairs cognition through the stresses of greater existential risk,[99] obesity itself imposes a cognitive cost, including deficits in episodic memory and decision-making.[100,101] The result may be a vicious cycle among Western diet, obesity, and cognitive disorders that inhibits how dietary impulses respond to caloric need.[102]

These questions lie at the origins and development of the inverse correlation between wealth and obesity. In wealthy countries, the phenomenon is known as the "reverse gradient in obesity," by which lower-income households tend to have higher rates of obesity.[103,104] Under one hypothesis, income enables people to afford expensive foods and the resources to exercise effectively.[103,105]

The reverse gradient does not exist in developing countries. Indeed, it is called "reverse gradient" because among the poorest 80% of society, more wealth means you can eat more, such that increased wealth correlates with increasing body weight.[105] Demographic and health-survey data from rural women in low-income countries show that household wealth is a predictor of mean body-mass index at the country level.[105] Above $400 per capita, BMI increases with both wealth and age.[105] Below this point, body-mass index becomes so low as to show little effect of either age or wealth.

The reverse gradient is a recent phenomenon in developed countries over the last several decades. Between 1990 and 2014, obesity in the United States is more than doubled, and in some states reached over a third of the population. The reverse gradient between wealth and obesity originated and intensified during that time. In Europe, the negative gradient may have emerged as recently as 2008.[106]

6. CONCLUSIONS

Sugar is a global industry integral to the American economy,[107] with centuries of complex history,[108] running from a notorious global commodity of transatlantic slave trade centuries ago[68,109] to a global commerce that contributes over 20 teaspoons of added sugar (over 30 for adolescent males) per day to the average US diet.[73] A simple policy approach, such as taxing sugar-sweetened beverages[74] is difficult after 50 years of government lobbying that has obscured the well-demonstrated harmful effects of sugar.[76]

These harmful effects include obesity, which parallels the aging process through the alteration of metabolic pathways. It appears that humans have entered an unprecedented era in terms of metabolism and aging for much of the population, whose expected lifespan, which now nears 80 years of age, may be complicated by the metabolism-altering conditions of diabetes and obesity. Because the consumption of refined sugar is strongly correlated with household economics, the heterogeneity of human aging and metabolism will remain a complex question of behavior and socioeconomic disparities.

REFERENCES

1. Flegal KM, Caroll MD, Ogden CL, Curtin LR. Prevalence and trends in obesity among US adults, 1999–2008. *JAMA*. 2010;303:235–241.
2. Ogden CL, Carroll MD, Kit BK, Flegal KM. Prevalence of childhood and adult obesity in the United States, 2011–2012. *JAMA*. 2014;311:806–814.
3. Colman RJ, Beasley TM, Kemnitz JW, Johnson SC, Weindruch R, Anderson RM. Caloric restriction reduces age-related and all-cause mortality in rhesus monkeys. *Nat Commun*. 2014;5:3557.
4. Basu S, Yoffe P, Hills N, Lustig RH. The relationship of sugar to population-level diabetes prevalence: an econometric analysis of repeated cross-sectional data. *PLoS One*. 2013;8(2):e57873.
5. Johnson RK, Appel LJ, Brands M, Howard BV, Lefevre M, Lustig RH, Steffen L, Wylie-Rosett J. Dietary sugars intake and cardiovascular health. *Circulation*. 2009;120:1011–1120.
6. Riera CE, Dillin A. Tipping the metabolic scales towards increased longevity in mammals. *Nat Cell Biol*. 2015;17:196–203.
7. Burkewitz K, Zhang Y, Mair WB. AMPK at the nexus of energetics and aging. *Cell Metab*. 2014;20:10–25.
8. Kanfi Y, Naiman S, Amir G, Peshti V, Zinman G, Nahum L, Bar-Joseph Z, Cohen HY. The sirtuin SIRT6 regulates lifespan in male mice. *Nature*. 2012;483:218–221.
9. Tomimatsu K, Narita M. Translating the effects of mTOR on secretory senescence. *Nat Cell Biol*. 2015;17:1230–1232.
10. Finkel T. The metabolic regulation of aging. *Nat Med*. 2015;21:1416–1423.
11. Madeo F, Zimmermann A, Maiuri MC, Kroemer G. Essential role for autophagy in lifespan extension. *J Clin Invest*. 2015;125:85–93.

12. Speakman JR, Mitchell SE. Caloric restriction. *Mol Aspects Med.* 2011;32:159–221.
13. Bartke A, Westbrook R. Metabolic characteristics of long-lived mice. *Front Genet.* 2012;3:288.
14. Liao CY, Rikke BA, Johnson TE, Gelfond JA, Diaz V, Nelson JF. Fat maintenance is a predictor of the murine lifespan response to dietary restriction. *Aging Cell.* 2011;10:629–639.
15. Ravussin E, Redman LM, Rochon J, Das SK, Fontana L, Kraus WE, Romashkan S, et al. A 2-year randomized controlled trial of human caloric restriction: feasibility and effects on predictors of health span and longevity. *J Gerontol A.* 2015;70:1097–1104.
16. Spurlock ME, Gabler NK. The development of porcine models of obesity and the metabolic syndrome. *J Nutr.* 2008;138:397–402.
17. Tardif SD, Power ML, Ross CN, Rutherford JN, Layne-Colon DG, Paulik MA. Characterization of obese phenotypes in a small nonhuman primate, the common marmoset (*Callithrix jacchus*). *Obesity.* 2009;17:1499–1505.
18. Vandamme TF. Use of rodents as models of human diseases. *J Pharm Bioallied Sci.* 2014;6:2–9.
19. Shively CA, Clarkson TB. The unique value of primate models in translational research. *Am J Primatol.* 2009;71:715–721.
20. Bodkin N, Hannah J, Ortmeyer H, Hansen B. Central obesity in rhesus monkeys: association with hyperinsulinemia, insulin resistance and hypertriglyceridemia? *Int J Obes.* 1993;17:53–61.
21. Comuzzie AG, Cole SA, Martin L, Carey KD, Mahaney MC, Blangero J, VandeBerg JL. The baboon as a nonhuman primate model for the study of the genetics of obesity. *Obes Res.* 2003;11:75–80.
22. Didier ES, MacLean AG, Mohan M, Didier PJ, Lackner AA, Kuroda MJ. Contributions of nonhuman primates to research on aging. *Vet Pathol.* 2016;53:277–290.
23. Finch C, Austad S. Primate models for human brain aging and neurological diseases. *Ann Rev Gerontol Geriatr.* 2014;34:139–170.
24. Hoffman CL, Higham JP, Mas-Rivera A, Ayala JE, Maestripieri D. Terminal investment and senescence in rhesus macaques (*Macaca mulatta*) on Cayo Santiago. *Behav Ecol.* 2010;21:972–978.
25. Ely JJ, Zavaskis T, Lammey ML. Hypertension increases with aging and obesity in chimpanzees (*Pan troglodytes*). *Zool Biol.* 2013;32:79–87.
26. Lowenstine LJ, McManamon R, Terio KA. Comparative pathology of aging great apes: bonobos, chimpanzees, gorillas, and orangutans. *Vet Pathol.* 2016;53:250–276.
27. Videan EN, Fritz J, Murphy J. Effects of aging on hematology and serum clinical chemistry in chimpanzees (*Pan troglodytes*). *Am J Primatol.* 2008;70:327–338.
28. Kemnitz JW. Calorie restriction and aging in nonhuman primates. *ILAR J.* 2011;52:66–77.
29. Smucny DA, Allison DB, Ingram DK, Roth GS, Kemnitz JW, Kohama SG, Lane MA. Changes in blood chemistry and hematology variables during aging in captive rhesus macaques (*Macaca mulatta*). *J Med Primatol.* 2001;30:161–173.
30. Tardif SD, Mansfield KG, Ratnam R, Ross CN, Ziegler TE. The marmoset as a model of aging and age-related disease. *ILAR J.* 2011;52:54–65.
31. Bagi CM, Volberg M, Moalli M, Shen V, Olson E, Hanson N, Berryman E, Andresen CJ. Age-related changes in marmoset trabecular and cortical bone and response to alendronate therapy resemble human bone physiology and architecture. *Anat Rec.* 2007;290:1005–1016.
32. Geula C, Nagykery N, Wu CK. Amyloid-β deposits in the cerebral cortex of the aged common marmoset (*Callithrix jacchus*): incidence and chemical composition. *Acta Neuropathol.* 2002;103:48–58.

33. Wu C-K, Nagykery N, Hersh LB, Scinto LFM, Geula C. Selective age-related loss of calbindin-D28k from basal forebrain cholinergic neurons in the common marmoset (*Callithrix jacchus*). *Neuroscience.* 2003;120:249–259.
34. Ross CN, Davis K, Dobek G, Tardif SD. Aging phenotypes of common marmosets (*Callithrix jacchus*). *J Aging Res.* 2012;567143.
35. Goryakin Y, Rocco L, Suhrcke M. The contribution of urbanization to non-communicable diseases: evidence from 173 countries from 1980 to 2008. *Econ Hum Biol.* 2017;26:151–163.
36. Jensen MD, Ryan DH, Donato KA, et al. Guidelines (2013) for managing overweight and obesity in adults. *Obesity.* 2014;22:S1–S410.
37. Brown JH, Gillooly JF, Allen AP, Savage VM, West GB. Toward a metabolic theory of ecology. *Ecology.* 2004;85:1771–1789.
38. Hudson LN, Isaac NJB, Reuman DC, Ardia D. The relationship between body mass and field metabolic rate among individual birds and mammals. *J Anim Ecol.* 2013;82:1009–1020.
39. Martin RD. Scaling of the mammalian brain: the maternal energy hypothesis. *Physiology.* 1996;11:149–156.
40. West GB, Brown JH, Enquist BJ. A general model for ontogenetic growth. *Nature.* 2001;413:628–631.
41. Gillooly JF, Charnov EL, West GB, Savage VM, Brown JH. Effects of size and temperature on developmental time. *Nature.* 2002;417:70–73.
42. Kaplan HS, Robson A. The emergence of humans: the coevolution of intelligence and longevity with intergenerational transfers. *Proc Nat Acad Sci USA.* 2002;99:10221–10226.
43. Del Giudice M, Gangestad SW, Kaplan HS. Life history theory and evolutionary psychology. In: Buss DM, ed. In: *Handbook of Evolutionary Psychology. New York, NY: Wiley; 2015:88–114.*
44. Williams GC. Pleiotropy, natural selection and the evolution of senescence. *Evolution.* 1957;11:398–411.
45. Gadgil M, Bossert WH. Life historical consequences of natural selection. *Am Nat.* 1970;104:1–24.
46. Kirkwood TBL, Austad SA. Why do we age? *Nature.* 2000;408:233–238.
47. Harpending H, Draper P, Pennington R. Cultural evolution, parental care and mortality. In: Swedlund AC, Armelgos GJ, eds. In: *Disease in Populations in Transition: Anthropological and Epidemiological Perspectives. New York, NY: Bergin and Garvey; 1990:251–265.*
48. Sear R. Evolutionary contributions to the study of human fertility. *Pop Stud.* 2015;69: S39–S55.
49. Kaplan HS, Hill K, Lancaster JB, Hurtado AM. A theory of human life history evolution: diet, intelligence, and longevity. *Evol Anthropol.* 2000;9:156–185.
50. Kaplan HS, Gangestad SW. Life history theory and evolutionary psychology. In: Buss DM, ed. In: *The Handbook of Evolutionary Psychology. Wiley: New York, NY; 2005:68–95.*
51. Cordain L, Eaton SB, Miller JB, Mann N, Hill K. The paradoxical nature of hunter-gatherer diets: meat-based, yet non-atherogenic. *Eur J Clin Nutr.* 2002;56:S42–S52.
52. Kaplan H, Thompson RC, Trumble BC, Wann LS, Allam AH, Beheim B, Frohlich B, et al. Coronary atherosclerosis in indigenous South American Tsimane: a cross-sectional cohort study. *The Lancet.* 2017;389(10080):1730–1739.
53. Mattson MP, Allison DB, Fontana L, Harvie M, Longo VD, Malaisse WJ, et al. Meal frequency and timing in health and disease. *Proc Natl Acad Sci USA.* 2014;111:16647–16653.

54. Catenacci VA, Pan Z, Ostendorf D, Brannon S, Gozansky WS, Mattson MP, Martin B, et al. A randomized pilot study comparing zero-calorie alternate-day fasting to daily caloric restriction in adults with obesity. *Obesity.* 2016;24:1874–1883.
55. Klempel MC, Kroeger CM, Varady KA. Alternate day fasting (ADF) with a high-fat diet produces similar weight loss and cardio-protection as ADF with a low-fat diet. *Metabolism.* 2013;62:137–143.
56. Kroeger CM, Klempel MC, Bhutani S, Trepanowski JF, Tangney CC, Varady KA. Improvement in coronary heart disease risk factors during an intermittent fasting/calorie restriction regimen: relationship to adipokine modulations. *Nutr Metabol.* 2012;9:98.
57. Wooding SP. Following the herd. *Nat Genet.* 2007;39:7–8.
58. Enattah NS, Sahi T, Savilahti E, Terwilliger JD, Peltonen L, Järvelä I. Identification of a variant associated with adult-type hypolactasia. *Nat Genet.* 2002;30:233–237.
59. Feldman MW, Cavalli-Sforzi LL. On the theory of evolution under genetic and cultural transmission with application to the lactose absorption problem. In: Feldman MW, ed. In: *Mathematical Evolutionary Theory. Princeton, NJ: Princeton University Press; 1989:145–173.*
60. Aoki KA. Stochastic model of gene–culture coevolution suggested by the "culture historical hypothesis" for the evolution of adult lactose absorption in humans. *Proc Nat Acad Sci USA.* 1996;83:2929–2933.
61. Bersaglieri T, Sabeti PC, Patterson N, Vanderploeg T, Schaffner SF, Drake JA, Rhodes M, Reich DE, Hirschhorn JN. Genetic signatures of strong recent positive selection at the lactase gene. *Am J Hum Genet.* 2004;74:1111–1120.
62. Gerbault P, Liebert A, Itan Y, Powell A, Currat M, Burger J, Swallow DM, Thomas MG. Evolution of lactase persistence: an example of human niche construction. *Phil Trans R Soc B.* 2011;366:863–877.
63. O'Brien MJ, Laland KN. Genes, culture and agriculture: an example of human niche construction. *Curr Anthropol.* 2012;53:434–470.
64. Bentley RA, O'Brien MJ, Manning K, Shennan S. On the relevance of the European Neolithic. *Antiquity.* 2015;89:1203–1210.
65. Cordain L, Eaton SB, Sebastian A, Mann N, Lindeberg S, Watkins BA, O'Keefe JH, Brand-Miller J. Origins and evolution of the Western diet: health implications for the 21st century. *Am J Clin Nutr.* 2005;81:341–354.
66. Guyenet S. By 2606, the US Diet will be 100 percent sugar. 2012. Available from: http://wholehealthsource.blogspot.com/2012/02/by-2606-us-diet-will-be-100-percent.html
67. White JS. Sucrose, HFCS, and fructose: history, manufacture, composition, applications, and production. In: Rippe JM, ed. In: *Fructose, High Fructose Corn Syrup, Sucrose and Health.. New York, NY: Springer; 2014:13–33.*
68. Vieira A. Sugar Islands: The sugar economy of Madeira and the Canaries, 1450–1650. In: Schwartz SB, ed. In: *Tropical Babylons: Sugar and the Making of the Atlantic World. Chapel Hill: University of North Carolina Press; 2004:42–84.*
69. Eltis D, Richardson D. The 'numbers game' and routes to slavery. *Slavery Abol.* 1997;18:1–15.
70. Avena NM, Rada P, Hoebel BG. Evidence for sugar addiction: behavioral and neurochemical effects of intermittent, excessive sugar intake. *Neurosci Biobehav R.* 2008;32:20–39.
71. Hu FB, Malik VS. Sugar-sweetened beverages and risk of obesity and type 2 diabetes: epidemiologic evidence. *Physiol Behav.* 2010;100:47–54.
72. Shang XW, Liu AL, Zhang Q, Hu XQ, Du SM, Ma J, Xu GF, et al. Report on childhood obesity in China (9): sugar-sweetened beverages consumption and obesity. *Biomed Environ Sci.* 2012;25:125–132.

73. Johnson RJ, Segal MS, Sautin Y, Nakagawa T, Feig DI, Kang DH, Gersch MS, Benner S, Sánchez-Lozada LG. Potential role of sugar (fructose) in the epidemic of hypertension, obesity and the metabolic syndrome, diabetes, kidney disease, and cardiovascular disease. *Am J Clin Nutr.* 2007;86:899–906.
74. Nakhimovsky SS, Feigl AB, Avila C, O'Sullivan G, Macgregor-Skinner E, Spranca M. Taxes on sugar-sweetened beverages to reduce overweight and obesity in middle-income countries: a systematic review. *PLoS One.* 2016;11(9):e0163358.
75. Lustig RH, Schmidt LA, Brindis CD. Public health: the toxic truth about sugar. *Nature.* 2012;482:27–29.
76. Kearns CE, Glantz SA, Schmidt LA. Sugar industry influence on the scientific agenda of the National Institute of Dental Research's 1971 national caries program: a historical analysis of internal documents. *PLoS Med.* 2015;12(3):e1001798.
77. Klimentidis YC, Beasley TM, Lin H-Y, Murati G, Glass GE, Guyton M, Newton W, et al. Canaries in the coal mine: a cross-species analysis of the plurality of obesity epidemics. *Proc R Soc B.* 2011;278:1626–1632.
78. Zoran DL. Obesity in dogs and cats: a metabolic and endocrine disorder. *Vet Clin North Am Small Anim Pract.* 2010;40:221–239.
79. Hernandez DC. The impact of cumulative family risks on various levels of food insecurity. *Soc Sci Res.* 2015;50:292–302.
80. Shrewsbury V, Wardle J. Socioeconomic status and adiposity in childhood: a systematic review of cross-sectional studies 1990–2005. *Obesity.* 2012;16:275–284.
81. Nettle D, Andrews C, Bateson M. Food insecurity as a driver of obesity in humans: the insurance hypothesis. *Behav Brain Sci.* 2017.
82. Song WO, Wang Y, Chung CE, Song B, Lee W, Chun OK. Is obesity development associated with dietary sugar intake in the U.S.? *Nutrition.* 2012;28:1137–1141.
83. Johnson ML, Distelmaier K, Lanza IR, Irving BA, Robinson MM, Konopka AR, Shulman GI, Nair KS. Mechanism by which caloric restriction improves insulin sensitivity in sedentary obese adults. *Diabetes.* 2016;65:74–84.
84. Bocarsly ME, Powell ES, Avena NM, Hoebel BG. High-fructose corn syrup causes characteristics of obesity in rats: increased body weight, body fat and triglyceride levels. *Pharmacol Biochem Behav.* 2010;97:101–106.
85. Bray GA, Nielsen SJ, Popkin BM. Consumption of high-fructose corn syrup in beverages may play a role in the epidemic of obesity. *Am J Clin Nutr.* 2004;79:537–543.
86. Jürgens H, Haass W, Castañeda TR, Schürmann A, Koebnick C, Dombrowski F, Otto B, et al. Consuming fructose-sweetened beverages increases body adiposity in mice. *Obes Res.* 2005;13:1146–1156.
87. Go Y-M, Liang Y, Uppal K, Soltow QA, Promislow DE, Wachtman LM, Jones DP. Metabolic characterization of the common marmoset (*Callithrix jacchus*). *PLoS One.* 2015;10:e0142916.
88. Redsell SA, Atkinson P, Nathan D, Niroshan Siriwardena A, Swift JA, Glazebrook C. Parents' beliefs about appropriate infant size, growth and feeding behaviour: implications for the prevention of childhood obesity. *BMC Pub Health.* 2010;10:7–11.
89. Ross CN, Power ML, Artavia JM, Tardif SD. Relation of food intake behaviors and obesity development in young common marmoset monkeys. *Obesity.* 2013;21:1891–1899.
90. Christakis NA, Fowler JH. Social contagion theory: examining dynamic social networks and human behavior. *Stat Med.* 2013;32:556–577.
91. Moreno JP, Vaughan E, Hernandez DC, Cameron RT, Foreyt JP, Johnston CA. The impact of acculturation level on weight status and weight outcomes in Hispanic children. *J Racial Ethn Health Disparities.* 2016;3:582–589.

92. Shakya HB, Christakis NA, Fowler JH. Self-comparisons as motivators for healthy behavior. *Obesity.* 2015;23:2477–2484.
93. So J, Prestin A, Lee L, Wang Y, Yen J, Chou WYS. What do people like to "share" about obesity? A content analysis of frequent retweets about obesity on Twitter. *Health Comm.* 2015;31:193–206.
94. Pilling LC, Atkins JL, Bowman K, Jones SE, Tyrrell J, Beaumont RN, Ruth KS, et al. Human longevity is influenced by many genetic variants: evidence from 75,000 UK Biobank participants. *Aging.* 2016;8:547–560.
95. Jackson P, Viehoff V. Reframing convenience food. *Appetite.* 2015;98:1–11.
96. Hernandez DC, Pressler E. Accumulation of childhood poverty on young adult overweight or obese status: race/ethnicity and gender disparities. *J Epidemiol Comm Health.* 2014;68:478–484.
97. Hernandez DC, Pressler E, Dorius C, Mitchell KS. Does family instability make girls fat? Gender differences between instability and weight. *J Marriage Fam.* 2014;76:175–190.
98. Jolliffe D. Overweight and poor? On the relationship between income and the body mass index. *Econ Hum Biol.* 2011;9:342–355.
99. Mani A, Mullainathan S, Shafir E, Zhao J. Poverty impedes cognitive function. *Science.* 2013;341:976–980.
100. Cheke LG, Simons JS, Clayton NS. Higher body mass index is associated with episodic memory deficits in young adults. *Q J Exp Psychol.* 2016;69:2305–2316.
101. Reinert KRS, Po'e EK, Barkin S. The relationship between executive function and obesity in children and adolescents: a systematic literature review. *J Obes.* 2013;6:1–10.
102. Kanoski SE, Davidson TL. Western diet consumption and cognitive impairment: links to hippocampal dysfunction and obesity. *Physiol Behav.* 2011;1:59–68.
103. Hruschka DJ. Do economic constraints on food choice make people fat? A critical review of two hypotheses for the poverty–obesity paradox. *Am J Hum Biol.* 2012;24:277–285.
104. Subramanian SV, Perkins JM, Özaltin E, Davey Smith G. Weight of nations: a socioeconomic analysis of women in low- to middle-income countries. *Am J Clin Nutr.* 2011;93:413–421.
105. Hruschka DJ, Hadley C, Brewis A. Disentangling basal and accumulated body mass for cross-population comparisons. *Am J Phys Anthropol.* 2014;153:542–550.
106. Villar JG, Quintana-Domeque C. Income and body mass index in Europe. *Econ Hum Biol.* 2009;7:73–83.
107. Nestle M. Food industry funding of nutrition research: the relevance of history for current debates. *JAMA Intern Med.* 2016;176:1685–1686.
108. Horton M, Bentley RA, Langton P. A history of sugar—the food nobody needs, but everyone craves. *Conversation.* 2015.
109. Fogel RW. *Without Consent or Contract: The Rise and Fall of American Slavery.* New York, NY: Norton; 1994.

CHAPTER FOUR

The Intersection of Curandismo and Western Medicine and Their Epidemiological Impact for Aging Mexican Americans

Pamela Anne Quiroz[1]
Center for Mexican American Studies, University of Houston, Houston, TX, United States
[1] Corresponding author. E-mail address: paquiroz@Central.UH.EDU

Contents

Abstract

The rising costs of health care in the United States continue to stimulate interest in alternative health care options among Americans. Western medicine is also beginning to pay more attention to these alternative practices and their utility for successful treatment of illness. Alternative health care practices have always been used by Hispanics whose mortality and morbidity profiles have stimulated interest from researchers. Though Hispanics occupy some of the lowest socioeconomic positions in the United States, they have higher life expectancy rates and comparable rates of morbidity for many health conditions relative to other racial/ethnic groups. This has been referred to as the "Hispanic health advantage" and the "Hispanic health paradox" [Turra CM, Elo IT. The impact of salmon bias on hispanic mortality advantage: new evidence from social security data. *Popul Res Policy Rev.* 2008;27:515. Franzini L, Fernandez-Esquer ME. Socioeconomic, cultural, and personal influences on health outcomes in low income Mexican-origin individuals in Texas. Soc Sci Med. 2004;59

Progress in Molecular Biology and Translational Science, Volume 155
ISSN 1877-1173
http://dx.doi.org/10.1016/bs.pmbts.2017.12.001

(8):1629–1646. George M. The Mexican-American health paradox: the collective influence of sociocultural factors on Hispanic health outcomes. *Discussions*. 2013;9(2):2–3. Gallo LC, Penedo FJ, Espinosa de los Monteros K, Argueles W. Resiliency in the face of disadvantage: do hispanic cultural characteristics protect health outcomes. *J Pers*. 2009;77(6):1707–1746. Turra CM, Goldman N. Socioeconomic differences in mortality among U.S. adults: insights into the Hispanic paradox. *J Gerontol*. 2007;62(3):184–192]. Even when controlling for socioeconomic status and education, life expectancy for Hispanics is greater than that of whites (CDIC, 2015). Debate about the causes of this paradox continues and some data indicates that the paradox is specific to older Hispanics (i.e., middle and older ages). Among Hispanic ethnic groups, this "paradox" has been most salient for Mexican Americans, therefore, this chapter focuses on Mexican Americans and the intersections of traditional health care practices among Mexican Americans (curandismo) with the practice of Western medicine to explore how they combine and conflict to impact aging Hispanics in three health areas: diabetes, menopause, and dementia.

1. INTRODUCTION

The rising costs of health care in the United States continue to stimulate interest in alternative health care options among Americans. Western medicine is also beginning to pay more attention to these alternative practices and their utility for successful treatment of illness. Alternative health care practices have always been used by Hispanics whose mortality and morbidity profile for several health conditions has been characterized as a "health paradox." This term refers to the fact that Hispanics occupy some of the lowest socioeconomic positions in the United States yet have the highest life expectancy rates and comparable rates of morbidity for several health conditions relative to other racial/ethnic groups (this has also been referred to as the "Hispanic health advantage").[1–4] Even when controlling for socioeconomic status and education, life expectancy for Hispanics is greater than that of whites.[4a] Debate about the causes of this paradox continues and some data indicates that the paradox is specific to older Hispanics (i.e., middle and older ages). Among Hispanic ethnic groups, this "paradox" has been most salient for Mexican Americans, therefore, this chapter focuses on Mexican Americans and the intersections of traditional health care practices among Mexican Americans (curandismo) with the practice of Western medicine to explore how they combine and conflict to impact aging Hispanics in three health areas: diabetes, menopause, and dementia.

2. BACKGROUND

Hispanics comprise 18% of the population of the United States with 65% of Hispanics of Mexican descent. And though the Hispanic population as a whole is regarded as a young population, significant aspects of aging among Mexican Americans are worth examining. As with other minority groups, the morbidity profile of Hispanics generally, and Mexican Americans specifically, reflects several variables, most notably, diet, genetics, and structural inequalities. Illnesses disproportionately associated with Mexican Americans include diabetes, obesity, asthma, and HIV. Among older Mexican Americans hypertension, cardiovascular and renal disease are also found. At the same time, Hispanics are the largest subgroup in the United States without health insurance, as language barriers, affordability of healthcare, lack of minority physicians, expectations of discrimination, and poverty contribute to the current health care practices by Hispanics (Department of Health and Human Services, 2010). In light of these circumstances, many Hispanic ethnic groups continue to rely upon traditional health care practices, such as *Curandismo. Curandismo* traces back to Aztec, Mayan, and Inca tribes and their beliefs of harmony between nature, spirit, and body. It consists of a set of rituals and prayers, herbal and other treatments used to address a variety of conditions among Hispanics. These treatments include those recognized by modern medicine, as well as those not emphasized in standard health care practice to include herbal treatments, vitamins, massage and rubs, prayers and rituals, religious paraphernalia, such as pictures of saints and holy water, candles, incense, sprays, and amulets.[5] Indeed, *Curandismo* is regarded as a holistic approach to the physiological, emotional, spiritual, and social conditions that confront Hispanics. *Curanderos*, people who have the power to cure, are relied upon to treat somatic and social psychological ailments. These practitioners also have specialties on which they focus: *Yerberolas*, botanical and herbal remedies; *Parteras* or midwives; *Sobadora/o* massage; *Espiritualisto/as*, psychic mediums; and *Senor/as* who read tarot cards. In addition to curandismo, Hispanics also rely upon western medicine.[6,7] Remedies of *curandismo* also rely on diet with particular foods encouraged to maintain balance in the body and regulate heat and cold. Imbalances are regarded as caused by physical or emotional fatigue, relationship to another person, or an alteration in temperature (e.g., walking barefoot on a cold floor). Many foods are classified as *hot* or *cold* (in essence, as opposed to temperature). Herbs are used to treat stomach and intestinal issues, sore muscles, weight loss, acne, arthritis, HIV, and diabetes, to name a few. Spices are regarded as having medicinal properties (cilantro, chocolate, cumin, oregano, cinnamon, and chamomile). Medical research has examined some of the traditional practices of curandismo and found some utility in the components of these treatments for patients. The following sections address some of the more commonly recognized epidemiological issues for Mexican Americans by the American Medical Association and the Center for Disease Control and how they are addressed using both contemporary western medicine and curandismo.

3. DIABETES

Diabetes presents the greatest disparities in morbidity and mortality between Hispanics and non-Hispanics with rates of diabetes increasing.[8,9] US Hispanics are shown to have higher rates of diabetes and obesity relative to Hispanics who do not live in the United States. In fact, lower mortality risks for US-born Hispanics advantage is specific to Mexican Americans and does not apply to Puerto Ricans or Cubans.

A study of the perceptions among Mexicans, who resided in California, Florida, Illinois, and Texas, attributed their diabetes to both biological factors and cultural beliefs. Emphasis by participants was placed on symptoms (fatigue) rather than biological characteristics of the disease (high blood glucose). Diabetes was also attributed to lifestyle factors, such as smoking, drinking alcohol, and lack of rest. Some Mexican Americans also regarded the disease as punishment for these activities citing *Susto* as a factor in contracting diabetes. *Susto* is a folk illness caused by a traumatic event that causes the loss of soul. Symptoms can include insomnia, anxiety, loss of appetite, and social withdrawal.[9a] *Susto* is seen as causing the body to become more susceptible to diabetes. Mexican Americans pointed to susto combined with unhealthy lifestyle and heredity as the causes of diabetes; and it is difficult to know how these factors were weighted by participants. One study found that positive (extreme) emotional states were also seen as responsible for the onset of diabetes, such anger, sadness, or even joy.[10]

Though most studies on the use of alternative therapies for diabetes among Mexican Americans and Hispanics are qualitative studies, they suggest the common use of herbal treatments even though these therapies are not substituted for professional medical treatments. Among these therapies is the use of nopal (prickly pear cactus), aloe vera, and nispero (Chinese plum). The most commonly used alternative treatment is nopal. Though nopal has been used by Mexicans for therapeutic purposes since the time of the Aztecs, its use for treatment of diabetes did not occur until the 1990s. Nopal was prepared in a variety of ways: sautéed with other vegetables (onions, parsley, cilantro), prepared in soups, as a shake, or in water (nopal water).

Some regional variation also resulted in use of different herbs or alternative treatments. For example, in Ohio, cat's claw, corn silk, aloe vera, nopal, tea made from sheep stool, and olive oil mixed with a spoonful of the individual's urine was used while in Chicago these remedies included boiled eggplant, lemon mixed with olive oil, and grapefruit juice or skins.[9a] Another study found the majority of herbal treatments for diabetes involved the use of teas comprised of Bricklebush, Buckthorn, Eucalyptus, Tree of Life, and Yellow bells. Not surprising, the vast majority of Mexican Americans learn these practices from friends and family members

Mexican Americans who use these alternative therapies were aware their doctors did not support them and were even concerned about combining traditional medical therapies with alternative therapies. They typically addressed this by adjusting their use in accordance with changes made to their prescription medication by their doctors.[11] When traditional diabetes medications were increased, use of alternative therapies was reduced. Alternative therapies used by Mexican Americans are common and normative as most people are aware of the use of these therapies by members of the community.

4. MENOPAUSE

Until the past few years, research on menopause has been divorced from women's reproductive and dietary histories and their interaction with cultural and biological factors across the lifespan. Clearly, cultural practices have an impact on health outcomes and the dearth of research on Mexican American women; and menopause adds to the need to understand the connection between cultural practices, biological predispositions, socioeconomic indicators, delivery of health care services, and health care disparities.

Though relatively little is known about the impact of menopause on Hispanic (and Mexican) women, there is some data to suggest that physiological experiences are described as similar to those of non-Hispanic women. One study found that Hispanic women rarely discuss menopause with family members but they perceive this period and their symptoms to be positive.[12] Such nondisclosure may be linked to cultural attitudes and the use of health care providers and available opportunities to address the symptoms of menopause. Another study found that Mexican women viewed menopause as inevitable and a natural transition in the life course; however, women also desired support and information from medical doctors.[13] Qualitative data suggest that women typically seek relief from alternative therapies prior to seeking support from traditional medicine. Women seek relief from teas and herbal therapies and are less likely to seek out traditional medical support in part because of an "agenda dissonance" between what women want and expect from doctors and what doctors assume and expect from Mexican (and Hispanic) women.[14] Hispanic women expect patient-centered and individualized care and yet are reluctant to share information with providers. And when dissatisfied with provider care, women shift providers.

Other indicators of Mexican menopausal women's disparities in health care include diet, social class and education levels, place of residence, exercise, expectations, and the use of alternative therapies. One study found less commitment to physical activity by Latinas, as interactive with symptoms of menopause. Another indicates a sense of impotence among women as reason for procrastination and avoidance: there is nothing they can do about their symptoms and the health care system, which is difficult to access. Another aspect of procrastination (in seeking traditional medical support) was placing the needs of family ahead of their own needs, consistent with the concept of "Marianismo." Marianismo refers to a cultural expectation tied to religion that directs women to be pure, nurturing, religious, and suffer.[15] Alternative therapies combined with support from families may mitigate the need for medical interventions.

5. DEMENTIA

Dementia is an umbrella term for a group of cognitive disorders that characterize memory impairment and difficulties with cognitive function. The literature demonstrates positive health outcomes for Hispanic elderly and substantial support of care for the elderly among Mexican Americans, even when dementia is present. This value of familism typically results in the reduced institutionalization of Hispanic elderly. Another mechanism used is adult day care, which allows Hispanic families to keep parents at home. Acceptance of health and dependence on family is directly related to responses to health problems.[16] At the same time, such caregiving and adherence to cultural expectations can also generate stress for family members precisely because there is minimal reliance on external resources. Use of external resources is a function of multiple factors, including familism and also lack of knowledge about services, language limitations, expectations of discrimination, and lack of health insurance.[17] US Latinos are less likely to seek healthcare than are other racial/ethnic groups in the United States. Though Mexican Americans do seek professional healthcare, curandismo is also relied upon by family members (caretakers) and patients as part of humanistic and spiritual support, and inclusion of family in treatment if not healing. In addition to standard medications, teas, soups, rubs, prayers, and massage are part of the care of Mexican elders. One study focused on how older Mexican Americans viewed ageing found that to participants aging well-meant remaining "right in the mind." The link between aging and *espiritismo* (belief that good/evil spirits affect well-being) was revealed while physical impairments and social circumstances and involvement were absent. Curandismo was part of the perception that illness, or not being "right in the mind" is punishment for a sin (a fundamental belief in curandismo). Another term for this is susto. Broader in meaning than simply dementia, susto includes the perception of loss of soul that results in mental illness. It is the belief that the soul has left the body because of a traumatic event. In the case of susto, the soul is afraid to return to the body.

Studies in Texas have found that 75% of Mexican Americans use curandismo and complementary therapies integrated with western medicine, however, the majority of Mexican Americans (69%) do not inform their health care providers.[18]

It is important to understand that cultural values alone do not determine the health care decisions of Mexican Americans and access is just as important. Additionally, research finds that opportunities, environmental conditions, social, and economic status each contribute to health care adaptations by Hispanics in general and Mexican Americans specifically. Nevertheless, cultural practices and values are part of the hybrid nature of health care practices of Mexican Americans, and curandismo can be viewed as a part of the innovative "culturally competent" health practices of the Mexican American community that may contribute significantly to the health advantages seen among lower SES Mexican Americans. Researchers predict that assimilation will not alter the Mexican American community's engagement with this set of beliefs and practices. It will simply adapt them. As practitioners have pointed out offering folk care can draw individuals into professional health care. One such example is the "nurse-curandera," a bilingual nurse trained in folk healing and combining homeopathic healing with allopathic healing remedies. Nurse-curanderas provide the link between the community and western medicine with advice about the safety and utility of folk remedies for particular health conditions.

Despite the health advantage, there is a greater prevalence of poverty and chronic disease among older Hispanics than among all older persons.[19,20] Stanford's Geriatric department offers a culturally grounded blend of western medical interventions with curandismo described as Complementary and Alternative Medicine (CAM) and Healers: familismo, personalismo, presentismo, jerarquismo, and espiritismo. Not only Mexican Americans or Hispanics use alternative medicine, estimates are that 1 in 3 persons in the United States also use alternative medicine; particularly for chronic illness.[21] Use of herbal remedies is highest among women, Hispanics, and non-Hispanic ethnic minorities. The extent to which these practices are efficacious remains unknown, as there are few systematic studies that investigate their impact. We know that healthful aging, at least among Mexican Americans, is perceived not only as the absence of disease. It is also how disease or health conditions are perceived and addressed along with whether persons are embedded in networks of support. Cultural explanations and practices can be both positive and disadvantageous. The role of medical research is to investigate these practices, and support those that work to minimize the costs of health care disparities to both Mexican Americans and the US health care system. Many Hispanics have spent their lives working in low paying positions, often in harsh and dangerous jobs for very low pay. Many, if not most, have inadequate preventative health care. These factors often result in serious health problems later in life and the lack of financial resources necessary to maintain good health (Brown, 2006).[22] Family is often the primary source of help for aging Latinos, and curandismo. Research supports the utility of cultural identity and strong social networks in fostering or protecting good health for a subset of Mexican Americans. Still, it is clear that income and wealth are generally positively connected to good health, and the lack of resources presents a major public health problem for significant numbers of not only Mexican Americans but also members of other racial/ethnic groups (African Americans, Puerto Ricans). Latinos face greater risk due to limited community-based support services, lack of bilingual health care professionals, and low quality of treatment and preventative care. As these problems have become the problems of increasing numbers of "forgotten Americans," research on health care disparities should involve systematic investigations of how cultural identity and alternative health care practices intersect with western medicine.

REFERENCES

1. Turra CM, Elo IT. The impact of salmon bias on hispanic mortality advantage: new evidence from social security data. *Popul Res Policy Rev.* 2008;27:515.

1a. Franzini L, Fernandez-Esquer ME. Socioeconomic, cultural, and personal influences on health outcomes in low income Mexican-origin individuals in Texas. *Soc Sci Med.* 2004;59(8):1629–1646.

2. George M. The Mexican-American health paradox: the collective influence of sociocultural factors on hispanic health outcomes. *Discussions.* 2013;9(2):2–3.

3. Gallo LC, Penedo FJ, Espinosa de los Monteros K, Argueles W. Resiliency in the face of disadvantage: do Hispanic cultural characteristics protect health outcomes. *J Pers.* 2009;77(6):1707–1746.

4. Turra CM, Goldman N. Socioeconomic differences in mortality among U.S. adults: insights into the Hispanic paradox. *J Gerontol.* 2007;62(3):184–192.

4a. CDIC.Vital Signs: Leading Causes of Death, Prevalence of Diseases and Risk Factors, and Use of Health Services Among Hispanics in the United States 2009–2013. *MMWR Morb Mortal Wkly Rep.* 2015;64(17):469–478.

5. Ortiz B, Shields KM, Clauson KA, Clay PG. Complementary and alternative medicine use among Hispanics in the United States. *Ann Pharmocother.* 2007;41:994–1004.

6. Torres E, Sawyer T. *Curandero: A Life in Mexican Folk Healing.* Albuquerque, NM: University of New Mexico Press; 2005.

7. Applewhite SL. Curanderismo: demystifying the health beliefs and practices of elderly Mexican Americans. *Health Soc Work.* 1995;20(4):247–253.

8. Salinas JJ, Abdelbary BE, Rocha EA, Al Snih Z. Contextualizing the burden of chronic disease: diabetes, mortality and disability in older Americans. In: Angel J, Torres-Gil F, Markides K, eds. In: *Aging, Health and Longevity in the Mexican Origin Population. Boston, MA: Springer Publishers; 2012:145–158.*

9. Vega WA, Rodriguez MC, Gruskin E. Health disparities in the Latino population. *Epidemiol Rev.* 2009;31:99–112.

9a. Caban A, Walker E. A systematic review of research on culturally relevant issues for hispanics with diabetes. *Diabetes Educ.* 2006;32(4):584–595.

10. Coronado GD, Thompson B, Tejeda S, Godina R. Attitudes and beliefs among Mexican Americans about type 2 diabetes. *J Health Care Poor Underserved.* 2004;15:576–588.

11. Poss JE, Jezewski MA, Stuart AG. Home remedies for type 2 diabetes used by Mexican Americans in El Paso, Texas. *Clin Nurs Res.* 2007;12(4):304–323.

12. Zenk SN, Shaver JLF, Peragallo N, Fox P, Chavez N. Use of herbal therapies among midlife Mexican women. *Health Care Women Int.* 2010;22(6):585–597.

13. Melby M, Lock M, Kaufert P. Culture and symptom reporting at menopause. *Hum Reprod Update.* 2006;11(5):495–512.

14. Esposito N. Agenda dissonance. *Clin Nurs Res.* 2005;14(1):32–56.

15. Garces IC, Scarinci IC, Harrison L. An examination of sociocultural factors associated with health and health care seeking among Latina immigrants. *J Immigr Minor Health.* 2006;8(4):377–385.

16. Padilla YC, Villalobos G. Cultural responses to health among Mexican American women and their families. *Fam Community Health.* 2007;30(1):24–33.

17. Lopez R. Indigenous health and coping resources in Mexican American communities. In: Caldera YM, Lindsey E, eds. In: *Mexican American Children and Families. New York, NY: Routledge; 2014:212–225.*
18. Tafur M, Crowe T, Torres E. A Review of curandismo and healing practices among Mexicans and Mexican Americans. *Occup Ther Int.* 2009;16(1).82–88.
19. Palloni A. Health status of elderly Hispanics in the United States. In: Angel JL, Whitfield KE, eds. In: *The Health of Aging Hispanics: The Mexican-Origin Population. Springer Publishers; 2007:17–25.*
20. Sharkey J, Sharf B, St. John JA. Una persona derechita (staying right in the mind): perceptions of Spanish-spekaing Mexican American older adults in South Texas Colonias. *Gerontologist.* 2009;49(51):579–585.
21. Periyakoil VJ. *eCampus Geriatrics.* 2010. Available from: http://geriatrics.stanford.edu
22. Kolb P. *Social Work Practice with Ethnically and Racially Diverse Nursing Home Residents and Their Families.* New York, NY: Columbia University Press; 2010.

FURTHER READING

Angel J, Angel R. Minority group status and healthful aging: social structure still matters. *Am J Public Health.* 2006;96(7):1152–1159.

Brown S. Considering *Curandismo*: The Place of Traditional Hispanic Folk Healing in Modern Medicine. 2008. Available from: http://www.bc.edu/clubs/mendel/ethos/archives/2008/brown.shtml

Arlene. A systematic review of research on culturally relevant issues for Hispanics with diabetes. *Diabetes Educ.* 2006;12(4):584.

Center for Disease Control and Prevention (CDCP). Hispanic Health. 2015. Available from: https://www.cdc.gov/vitalsigns/hispanic-health/index.html

Yaffe KY, Haan M, Blackwell T, Cherkasova E, Whitmore R, West N. Metabolic syndrome and cognitive decline in elderly Latinos: findings from the Sacramento Area Latino Study of Ageing Study. *JAGS.* 2007;55(3):758–762.

CHAPTER FIVE

The Necessity of Active Muscle Metabolism for Healthy Aging: Muscular Activity Throughout the Entire Day

Marc T. Hamilton[1], Deborah G. Hamilton, Theodore W. Zderic
Texas Obesity Research Center and Health and Human Performance, University of Houston, Houston, TX, United States
[1] Corresponding author. E-mail address: mhamilton7@uh.edu

Contents

Progress in Molecular Biology and Translational Science, Volume 155
ISSN 1877-1173
http://dx.doi.org/10.1016/bs.pmbts.2017.12.014

Abstract

There is more need for "a movement–movement" than ever before. The percentage of seniors in our population is rising exponentially. Sedentary lifestyles throughout the lifespan have become the norm, including inactive youth and a sedentary workforce. Preventable chronic diseases caused by sedentary living have both lowered the quality of life for those directly affected or their families, and have created an unsustainable economic dilemma. In this article, we explain that whether it is a sedentary student, worker, or retiree, the most neglected but essential facts are as follows. By far, the most potent and rapid way to raise the rate of healthy metabolic and cardiovascular processes is through the immediate benefits of muscle contractions. Working muscle demands more energy and fuel than any other tissue in the body, but during inactivity the metabolic rate of muscle is relatively low. Depending on the type of contraction, muscle type, and other factors, the local fuel requirements within the working muscle can help to manage metabolic risks through a variety of processes, such as blood glucose utilization, uptake of unhealthy blood triglycerides, and increased blood flow. Given the large amount of time that people spend inactive each day, there is an enormous opportunity to raise the bar in optimizing health throughout the entire lifespan. Developed correctly, safe and low effort muscular activity can be performed for relatively long periods of time each day by the elderly and all segments of the population to optimize health and well being during aging.

1. SEDENTARY TIME EQUALS MUSCULAR INACTIVITY

From a physiological perspective, sedentary time is defined by a state of muscular inactivity, resulting in a low metabolic rate in muscle. When not sedentary, the energy expenditure of the body rises because of a *local* increase in the metabolic rate within the working muscle mass. It is the increased demand for fuel in the muscle fibers that are recruited during non-sedentary periods of the day that drives the metabolic responses and blood flow. People in the modern world must sit for many hours throughout each day. Thus, operationally in the behavioral literature, sedentary time has generally been described as the time people spend sitting doing inactive behaviors,[1] and thus not producing powerful enough muscle contractions and staying close to resting metabolic rate. Our whole room calorimetry study has confirmed that behaviors, such as watching TV, using a computer, and reading throughout much of the day result in similar energy expenditures that are close to resting metabolic rate.[2] The consensus from basic science studies has revealed that there is a growing list of unhealthy biochemical and physiological responses originating from inactive skeletal muscle.[3–12]

In this article, we provide the key concepts for understanding why a high amount of muscular inactivity (i.e., during sedentary behaviors) throughout the lifetime is a ubiquitous hazard to human health. This inactive modern lifestyle spans throughout most of the ordinary workday, and then is particularly problematic as we age in retirement. But we believe solutions are soon on the horizon. This article serves as a guide for developing a scientifically innovative and effective solution that is rooted in a more clear understanding about how the body works and controls metabolic processes.

Key concepts for understanding why inactivity is hazardous to human health are presented below. These concepts are intended to serve as a guide in designing efficacious solutions to the problems caused by inactivity in all ages of our population.

2. THE GOAL IS TO REPLACE AS MUCH SEDENTARY TIME AS POSSIBLE WITH AN EQUAL AMOUNT OF SAFE AND EFFECTIVE MUSCULAR ACTIVITY THROUGHOUT THE DAY

Leaders have frequently focused on the need for increasing activity of elderly people or on creating a more active workplace. Below we have summarized the rationale for these goals.

2.1 For Individuals to Age Optimally

From a metabolic standpoint, much of what we tend to associate with aging is largely preventable and not due to an intractable aging process per se. People, regardless of genetics, are looking for a more convenient and effective way to stay healthy and productive as they age. A personalized lifestyle strategy is needed that evolves over time with individuals as they age. That has to include expertise and innovative breakthroughs about how to maximize the benefits of muscular activity in an elderly population that is increasingly unable to engage in traditional exercise or physical activity.

2.2 For Workplace Wellness and Other Programs to be Successful in Preventing Unnecessary Disease

Self-insured corporations are especially looking for the benefits of creating an active workplace that reduces health care costs and improves worker productivity. Although it is hard to estimate, hundreds of billions are spent in the United Stated for self-insured healthcare costs, plus billions on corporate wellness programs that have arguably not been effective at making people nearly active enough.

3. WE HAVE REACHED A POINT IN HUMAN HISTORY WHERE THE UBIQUITOUS ACT OF SITTING IN A CHAIR HAS SURPASSED SLEEPING AS THE MOST ABUNDANT BEHAVIOR

Objective evidence (not self-reported) from wearable devices suggests that sitting time averages about 9–10 h/day.[13–17] Ironically, although prolonged sitting is ubiquitous and abundant, there has historically been far more research about the effects of 2.5 h per week of moderate physical activity in 10 min bouts of movement, that less than 10% of the public participates in regularly.[18,19] Additionally, perhaps even more importantly, for the person who is sitting inactive for ~70 h or more a week, that 2.5 h is hardly a dent in the overall pattern of muscular inactivity throughout the whole day. The goals for public health gains should be focused on the development of an innovative novel approach for all kinds of people to achieve the metabolic and cardiovascular benefits of muscular activity, as much as they can.

4. THERE ARE MANY DETRIMENTAL EFFECTS DIRECTLY CAUSED BY MANY HOURS PER DAY OF MUSCULAR INACTIVITY, REGARDLESS OF AGE

Two examples are for a lowering of lipoprotein lipase (LPL) activity with ensuing metabolic problems,[4,6] and deep vein thrombosis (DVT). Both are frequently a concern during immobility conditions when muscles in the legs are relatively inactive.[9,20] Simply put, it is important to understand that although some conditions are frequently linked to poor aging, they are also impacted by rather short periods of inactivity in supposedly healthy younger people.

5. THE METABOLIC RATE OF MUSCLE IS DETERMINED BY WHETHER OR NOT THERE ARE ONGOING ACUTE PERIODS OF CONTRACTILE ACTIVITY OR MUSCULAR INACTIVITY, REGARDLESS OF AGE

Although the elderly may have reduced total body metabolic rate at rest by a variety of processes, it is an incontrovertible fact that once a muscle is contracting (even during low effort behaviors), there is nothing that creates more potential to instantly raise metabolic rate. There is no pharmaceutical, nutritional trick, or genetic influence that comes close to quickly and profoundly raising metabolic rate more than periods of physical activity. Even low intensity physical activity at 1.5 METs (a 50% rise in metabolic rate from resting) is more than any other known stimulus of which we are aware.

6. WHEN ACTIVE, MUSCLE IN ELDERLY PEOPLE OR ANIMALS RETAINS ITS ABILITY TO DISPLAY HEALTHY METABOLIC RESPONSES

By and large, it is safe to conclude that the health promoting aspects of a high muscle metabolism do not deteriorate during aging if the muscle is heavily used. Elderly people retain the ability to have a rapid rise in the rate of glucose and fat utilization during acute periods of activity.[21] Those and other related processes, such as muscle blood flow are more a function of how much time a muscle is used (i.e., how many hours per day) than anything else of which we are aware, including age. But if muscle becomes inactive, regardless of age, it is sure to deteriorate in many ways, as well as have far reaching systemic metabolic consequences. Muscle in old age is still able to undergo favorable mitochondrial biogenesis (growing more mitochondria)[22,23] and angiogenesis (growing blood vessels)[24] essential for fatigue resistance and optimal metabolism. Thus, it is often believed that the "effects of aging" on many aspects of metabolic control and health are not truly effects of aging per se, but rather the effects of inactivity.

7. THE TENETS OF INACTIVITY PHYSIOLOGY FOR PROMOTING HEALTHY AGING

Inactivity physiology is a field of study we began in the early 2000s that has been emerging exponentially especially since ~2007; there are seven tenets comprising its foundation.[6,8,25] It is vital to understand these tenets completely in order to appreciate the magnitude of the impact any future public health solution will have that is born from this novel approach. The tenets listed below are excerpts from another article of ours.[25] They underscore the importance for understanding why sedentary time (i.e., inactive muscles throughout too much of the day) is one of the more ubiquitous health hazards of our modern age, especially in elderly people.

Tenet 1: Sedentary time is surprisingly hazardous to human health in many ways. We began to see the large and rapid effects of muscular inactivity on lipid metabolism[4] and diverse effects on gene expression for many metabolic and other processes[5] in basic pre-clinical studies. It followed, then, that there may be significance to those findings as it relates to people during inactive sitting as reviewed elsewhere.[6,8,11,25–27] The exposure rate to the risk of sedentary time is high because people of all ages and demographics sit or are not moving for a large amount of time each day.[13–17,27–30] Because of this, the majority of people are in serious need of a healthy active lifestyle more than previously realized. That said, once an effective behavioral solution is discovered, it will be a genuine novel breakthrough capable of creating enormous positive impact.

Tenet 2: Sedentary time is a distinct behavioral class from the time spent engaged in traditional MVPA recommendations. This tenet can be paraphrased simply as the time we spend doing traditional exercises does not reduce total daily sedentary time or the number of sitting periods.[13,31,32] More technically, this tenet raised the hypothesis that the various times people spend sitting inactive or participating in moderate-vigorous physical activity (MVPA) based leisure-time physical activity are distinct and totally different classes of behavior, with distinct determinants and independent effects on risk for disease.[13,31,32] Adding more 30 min of physical activity on 5 days/week (i.e., 2.5 h/week), in the form of MVPA in bouts of at least 10 min, has been widely recommended.[18] Considering that 30 min is only 1/48th of the day and people are often sitting for over 70 h/week, it should not be surprising that the recommended amount of MVPA hardly makes a dent in the massive amount of sedentary time. A simple point that is seemingly lost by many and causing conceptual problems related to tenet 2, is the fact that physical activity exists on a continuum and the most abundant type of physical activity is low intensity physical activity (LIPA). Logically, therefore inactivity is the absence of activity, not just the absence of one type of activity that is less frequent and more intense (MVPA).

Tenet 3: The specificity principle—the molecular stimuli and physiological benefits invoked by muscular activity are specific to the dose and type of activity. This is arguably the most fundamental concept needed to understand how to best perform muscular activity for obtaining the desired responses. By linking the specificity principle with a good understanding of physiology, it is possible to begin to understand the immediate benefits of muscular movement throughout the day, as well as the adaptive responses accumulating over time. This principle is critical for developing high quality studies to determine dose-response relationships. Approaching physiology studies with a firm grasp of the specificity principle (a foundational principle of exercise physiology), we started to see that for some key cellular and molecular factors important for disease risks,[3–5,33] it was necessary to distinguish the responses of inactivity physiology versus classical exercise physiology. Addtionally, we are probably still at the early stage for seeing how surprisingly necessary it is to have a high frequency and high duration of muscular activity throughout the day; a large amount of muscular activity promotes optimal regulation of diverse categories of genes and proteins ranging from lipid regulation by LPL, genes involved in the regulation of inflammation, glucose regulation, and homeostasis, etc. Most typically, people sit down about 50–60 times per day, and the cumulative duration generally averages 60 or more hours per week.[14] Certainly, by 2008,[8] and 2014,[11] we had begun to propose that it was implausible to think that inserting a bolus of MVPA-style exercise in 1–4 sessions per week would invoke the same cellular stimuli as hundreds of bouts of muscular activity spread throughout the entire day. The immediate benefits of muscular activity must surely be evident more frequently by combating sedentary time (tenet 5). And the underlying causes for the adaptive effects would also likely be from very different dose response stimuli. Undoubtedly, the optimum intensity and mode for high durations of physical activity is dependent in part upon fatigue, feasibility, safety, and physiological effectiveness. We elaborate more fully on this concept in a previous publication.[25]

Tenet 4: In people who do not already exercise, it logically follows that further increased rates for disease and mortality cannot be caused by additional exercise deficiency. The potential benefits of replacing large and frequent amounts of sedentary time with safe and achievable physical activity should be widely discussed, without delay. It is of high significance to focus on viable solutions for the problems caused by inactivity in the most sedentary people, elderly included. The shape of the dose response curve between fitness/activity and mortality or disease outcomes has long suggested that the major public health benefits are by getting the most inactive people to move more rather than making average people exercise.[6] And if the ability to effectively replace sedentary time with a large amount of safe muscular activity is *not* developed, the trajectory for growing numbers of sedentary people with multiple chronic diseases will continue to escalate in the wrong direction. Technological advances and other causes promoting sedentary behavior seem to continually make growing numbers of people sit more. Thus, society may not have yet reached the pinnacle of seeing the full range of poor health caused by the impact of sedentary time on preventable conditions (such as type 2 diabetes, cardiovascular disease, obesity, metabolic syndrome, DVT, fatty liver disease, dementia, and some cancers).

Tenet 5: The rapid time course of health related responses to physical inactivity/activity is in large part because of the direct effects caused by muscle contractile inactivity/activity throughout the day. Nothing else than contractile inactivity/activity comes close to having a greater instantaneous effect on variations in metabolic rate throughout each hour of the day. Skeletal muscle undergoes greater swings in metabolic rate from moment to moment than any other tissue type in the body. The metabolic rate of a muscle fiber at rest is quite low, but can rise 50- to 100-fold immediately upon activation. As soon as a muscle starts contracting, ATP (the cellular energy currency) demand rises immediately by a large amount. Blood flow and other support processes can also change at the onset and cessation of contractile activity in order to support the energy demand of the working fibers. As one of several examples, these factors help to explain why after experimentally decreasing normal spontaneous ambulatory contractile activity, our laboratory found rapid and large changes in LPL activity in tandem with slower extraction of blood triglyceride locally in oxidative muscle, and significant decreases in HDL cholesterol.[4] These findings were true regardless of gender. That is likely only the foreshadowing of many other responses where LPL may play a role in explaining the phenotypes caused by sitting inactive or high amounts of low effort contractile activity, since low LPL is associated with several phenotypes important for disease risk. Acute slow walking breaks from sedentary time improved hemostatic factors[34] and lowered postprandial glucose and insulin[35] in overweight and obese individuals who otherwise sat for several hours in the postprandial period. Even in young, relatively fit and nonobese adults, a high duration of low intensity movement throughout the day led to a significantly greater rate of glucose uptake and insulin action.[36] Additional examples supporting this tenet are outlined in more detail in one of our recent publications.[25] For these reasons, it was written[25] " . . . public health campaigns should eventually provide people with a rich sense of 'metabolic awareness' about the concept that 'the body constantly senses and responds to movement vs inactivity', in tandem with behavioral strategies promoting the ability to take full advantage of every opportunity to move more. And this is good news likely for all kinds of people, because evidence indicates that the immediate benefits of movement are independent of other more intractable long-term issues like weight loss and cardiorespiratory fitness."

Tenet 6: The role of local contractile activity and muscle specificity. One of the most useful concepts in muscle physiology is that once a motor neuron is stimulated, the excitation of the affected muscle fibers responds with close to an all or none type of contractile phenomenon. Because of this, the force output within large weight bearing muscles of the leg are graded mostly by recruitment of more motor units, not by a graded intensity within the individual fibers. This fundamental physiology regarding how the body works when we move helps us to understand why "light" effort physical activity can produce powerful changes in cellular signaling, metabolism, and health outcomes. At the level of the fibers contributing to low effort activity, such as strolling during indoor walking, the rise in energy demand, and fuel supply to the working muscle cells is far from being a light stimulus inside the working cells. Phenotypic diversity between different fiber types means that all muscles are not created equal, nor do they respond similarly to physical activity. There is heterogeneity for the metabolic proteins, fatigue resistance, and the threshold for activation during graded contractile intensity. In each of our studies in rats, the local contractile activity within the slow-twitch oxidative muscle appears to be most important for explaining the reason for maintaining a high duration of daily low intensity physical activity in young adults[3,4,11] and with aging.[33] During aging, there is a local reduction in the LPL activity within the capillaries of the slow red oxidative soleus muscle, but not other muscle types that rely on glycolytic metabolism, such as the tibialis anterior.[33] Importantly, it is the soleus muscle that has also been shown to be the most responsive muscle in the body to physical inactivity as it relates to ordinary lifestyle movement. In fact, two strains of rats commonly used in aging research displayed a dramatic reduction in the soleus LPL activity, and this was by a posttranscriptional mechanism.[3,33] In young rats that become inactive experimentally, they display a similar reduction in soleus-specific LPL activity by the post transcriptional mechanism process.[4,7] This points to an inactivity-dependent process responsible for the lowering of muscle LPL. We previously reviewed[37] the effects of aging in humans and rats on how plasma lipoproteins, especially triglycerides, rise during aging in humans as a consequence of lowering LPL activity. In general, beginning in middle age, there is an elevation in the plasma triglycerides as a result of impairments in LPL-mediated TG clearance rates in older individuals. The fact that we had also found the LPL enzyme activity reduction was localized specifically within slow-twitch oxidative skeletal muscle, but not glycolytic muscle, is likely important to understand the underlying root causes. We hypothesize that the metabolic responses in relatively fatigue-resistant slow-twitch oxidative skeletal muscle have a dominant role in combating the diseases associated with too much sitting.

8. IN SUMMARY, THE *DIRECT CAUSES* OF UNHEALTHY VERSUS HEALTHY AGING WITH RESPECT TO METABOLISM ARE OFTEN ROOTED IN MUSCLE PHYSIOLOGY

The examples are too many to list in detail and thereby provide a sufficient bibliography here, but there are well-studied categories to direct future efforts with the most scientific impact and public health significance.

By far, the body tissue that can raise the metabolic rate the most in a person (regardless of age) is skeletal muscle when it is producing movement (i.e., acute contractile activity). It does not take long for this effect to generate many healthy metabolic responses (tenet 5). The metabolic rate within the working muscle fibers rises and falls over as much as 50- to 100-fold range during the moments when muscles are contracting or become inactive.

There are widespread health problems frequently evident during aging related to muscle metabolism. Taking glucose metabolism as an important case in point, skeletal muscle is responsible for the majority of the blood glucose utilization (over 80% of the plasma glucose disposal is through muscle glucose uptake), even at rest.[38] Insulin resistance is not only one of the earliest events in the pathology leading to diabetes, but insulin resistance is also at the root of metabolic syndrome and many of the aspects of poor aging because of compensatory hyperinsulinemia. High levels of insulin may lead to disturbances in everything from liver and brain function to many cardiovascular complications also associated with poor aging.[39–44] Fortunately, a potent solution for correcting insulin resistance at the level of the underlying molecular root causes is to raise muscular activity to a high degree. This would ideally be done with as much time as one can safely do each day without risking injury.

All muscles are not the same with respect to their biochemical properties or their responsiveness to inactivity/activity. Our laboratory has focused heavily on the soleus muscle during periods of inactivity versus prolonged and highly intermittent contractile activity throughout the day. One reason for studying this muscle is because it is the paragon for understanding how muscle biochemistry and blood flow regulation are vital for good health, which is linked to both the acute responses to physical activity, as well as the long term responses to aging. Much has been learned about the molecular and physiological aspects in this muscle that help to explain why optimal aging requires a large duration of daily contractile activity (as described more in tenet 6). This muscle is greatly impacted by both aging and inactivity in many ways. By virtue of a specialized anatomy, vasculature, and cellular properties, the soleus is the muscle in humans and rats that is richest in the red oxidative slow twitch muscle fibers that are capable of prolonged contractile activity.[45] This skeletal muscle is in the calf, but is in some ways analogous to cardiac muscle. Both are specialized for a large duration of contractile activity throughout the day because of supreme fatigue resistance. The circulation of blood through the body is accomplished not only by the pumping of blood from the heart, but is also facilitated by the difficult task of pumping blood "uphill" from the lower legs *and back to the heart*. By virtue of being the largest calf muscle in humans and also the most heavily recruited, the soleus is likely the main muscle responsible for the "muscle pump," which is a process for propelling venous blood in the lower extremity back to the heart.[46] The red oxidative slow twitch soleus working muscle is also the predominant site for triglyceride clearance from the blood,[3,4] lipid oxidation (fat burning) and controlling many health promoting metabolic pathways.[47] Multiple aspects of metabolic health are favorably impacted by enzymes, such as LPL and mitochondrial proteins. LPL is rich in chronically active soleus muscle.[3,4] We describe this enzyme most simply as the "vacuum cleaner for blood lipids," because it is the main protein responsible for the uptake and breakdown of plasma triglyceride. Given this background, it is important to notice that we found strong evidence[3] to support the hypothesis that LPL is regulated by local stimuli dependent on the duration of contractile activity (increased LPL) or contractile inactivity (decreased LPL), with transient daily effects in just a matter of several hours. This, in turn, impacted the working muscle's ability to refuel itself with fat (plasma triglyceride) for energy and contractile power. LPL is not only important for the clearance of triglyceride from the plasma, but also for high levels of the cardioprotective high density lipoprotein (HDL).[48] LPL is also the enzyme controlling production of ligands (i.e., stimulators) for a class of nuclear receptors (i.e., PPARs) that are often described as the master switches for controlling genes that regulate dozens of aspects about metabolism, including antiinflammatory effects[49] and mitochondrial biogenesis.[50] Taken together, we believe there is an abundance of evidence that some muscles are more richly endowed in metabolic properties for promoting health.

Tenet 7: We do not need a band-aid. We do need a revolutionary breakthrough! For a genuine public health solution to address the complete problem of sedentariness, a solution that is potent and safely attainable by most everyone is needed. Major advances moving forward can occur. But to get there and see the breakthroughs for what they are, we need to know what goals to set.

The seven tenets and concepts we have laid out previously,[25] and reiterate here, will hopefully point a way toward formulating those goals. There are three things that must be considered when searching for discoveries needed to find that solution scientifically, as well as to get buy-in from the public to adopt a major behavioral change. First, we cannot trivialize the process by aiming low or wasting limited resources on descriptive guesses about what to study without a good understanding for how the body works from translational insights. Second, we need to avoid the confusion in the public health messaging caused by pseudo-debates that appear too frequently in some areas of lifestyle medicine. Thus, we should work cooperatively and embrace solutions that raise activity by a large amount of the day and are safe and practical for the whole population, including the elderly, obese, and people with chronic conditions. A challenge of our time is how to make all types of people become much more active. This is especially important and challenging for the people who account for most of the healthcare costs and need an activity-based solution the most. Third, the experts and funders cannot be cynical, just because prior public health failures are many. History is filled with seismic shifts in the right direction for public health just when gloom and doom seemed everywhere.

9. CONCLUDING STATEMENTS

In this article, we have made the unambiguous argument for why the optimal lifestyle includes high duration and frequent muscular activity throughout the entire day. This article has provided some of the general principles we and others are using to ultimately develop the best solution for optimizing metabolism necessary for healthy aging. Taken together, we believe that this logical line of thinking points the way toward development of scientific innovation needed for breakthroughs in how to age optimally.

ACKNOWLEDGMENTS

We appreciate the support of the American Diabetes Association as we have continued to develop a roadmap for understanding how muscular activity may provide solutions for sedentary lifestyle diseases: grant # 1-15-TS-14.

REFERENCES

1. Tremblay MS, Aubert S, Barnes JD, Saunders TJ, Carson V, Latimer-Cheung AE, Chastin SFM, Altenburg TM, Chinapaw MJM. SBRN Terminology Consensus Project Participants.Sedentary behavior research network (SBRN): terminology consensus project process and outcome. *Int J Behav Nutr Phys Act.* 2017;14(1):75.
2. Newton Jr RL, Han H, Zderic T, Hamilton MT. The energy expenditure of sedentary behavior: a whole room calorimeter study. *PLoS One.* 2013;8(5):e63171.
3. Hamilton MT, Etienne J, McClure WC, Pavey BS, Holloway AK. Role of local contractile activity and muscle fiber type on LPL regulation during exercise. *Am J Physiol Endocrinol Metab.* 1998;275:E1016–E1022.
4. Bey L, Hamilton MT. Suppression of skeletal muscle lipoprotein lipase activity during physical inactivity: a molecular reason to maintain daily low-intensity activity. *J Physiol.* 2003;551(Pt 2):673–682.
5. Bey L, Akunuri N, Zhao P, Hoffman EP, Hamilton DG, Hamilton MT. Patterns of global gene expression in rat skeletal muscle during unloading and low-intensity ambulatory activity. *Physiol Genomics.* 2003;13(2):157–167.
6. Hamilton MT, Hamilton DG, Zderic TW. Exercise physiology versus inactivity physiology: an essential concept for understanding lipoprotein lipase regulation. *Exerc Sport Sci Rev.* 2004;32:161–166.
7. Zderic TW, Hamilton MT. Physical inactivity amplifies the sensitivity of skeletal muscle to the lipid-induced downregulation of lipoprotein lipase activity. *J Appl Physiol (1985).* 2006;100(1):249–257.
8. Hamilton MT, Hamilton DG, Zderic TW. Role of low energy expenditure and sitting in obesity, metabolic syndrome, type 2 diabetes, and cardiovascular disease. *Diabetes.* 2007;56:2655–2667.
9. Zderic TW, Hamilton MT. Identification of hemostatic genes expressed in human and rat leg muscles and a novel gene (LPP1/PAP2A) suppressed during prolonged physical inactivity (sitting). *Lipids Health Dis.* 2012;12:137.
10. Latouche C, Jowett JB, Carey AL, Bertovic DA, Owen N, Dunstan DW, Kingwell BA. Effects of breaking up prolonged sitting on skeletal muscle gene expression. *J Appl Physiol.* 2013;114(4):453–460.
11. Hamilton MT, Hamilton DG, Zderic TW. Sedentary behavior as a mediator of type 2 diabetes. *Med Sport Sci.* 2014;60:11–26.
12. Bergouignan A, Latouche C, Heywood S, Grace MS, Reddy-Luthmoodoo M, Natoli AK, Owen N, Dunstan DW, Kingwell BA. Frequent interruptions of sedentary time modulates contraction- and insulin-stimulated glucose uptake pathways in muscle: ancillary analysis from randomized clinical trials. *Sci Rep.* 2016;6: 32044.
13. Craft LL, Zderic TW, Gapstur SM, Vaniterson EH, Thomas DM, Siddique J, Hamilton MT. Evidence that women meeting physical activity guidelines do not sit less: an observational inclinometry study. *Int J Behav Nutr Phys Act.* 2012;9:122.

14. Van der Berg JD, Stehouwer CD, Bosma H, van der Velde JH, Willems PJ, Savelberg HH, Schram MT, Sep SJ, van der Kallen CJ, Henry RM, Dagnelie PC, Schaper NC, Koster A. Associations of total amount and patterns of sedentary behaviour with type 2 diabetes and the metabolic syndrome: The Maastricht Study. *Diabetologia.* 2016;59(4):709 718.
15. Bellettiere J, Winkler EAH, Chastin SFM, Kerr J, Owen N, Dunstan DW, Healy GN. Associations of sitting accumulation patterns with cardio-metabolic risk biomarkers in Australian adults. *PLoS One.* 2017;12(6):e0180119.
16. Biddle SJ, Edwardson CL, Wilmot EG, Yates T, Gorely T, Bodicoat DH, Ashra N, Khunti K, Nimmo MA, Davies MJ. A randomised controlled trial to reduce sedentary time in young adults at risk of type 2 diabetes mellitus: project STAND (Sedentary Time ANd Diabetes). *PLoS One.* 2015;10(12):e0143398.
17. Matthews CE, Keadle SK, Moore SC, Schoeller DS, Carroll RJ, Troiano RP, Sampson JN. Measurement of active and sedentary behavior in context of large epidemiologic studies. *Med Sci Sports Exerc.* 2018;50(2):266–276.
18. GovTrack.us. S. 2748—110th Congress (2008): Physical Activity Guidelines for Americans Act of 2008, GovTrack.us (database of federal legislation). Available from: www.govtrack.us/congress/bill.xpd?bill=s110-2748&tab=analysis
19. Tucker JM, Welk GJ, Beyler NK. Physical activity in U.S.: adults compliance with the Physical Activity Guidelines for Americans. *Am J Prev Med.* 2011;40(4):454–461.
20. Gathof BS, Picker SM, Rojo J. Epidemiology, etiology and diagnosis of venous thrombosis. *Eur J Med Res.* 2004;9(3):95–103.
21. Prior SJ, Ryan AS, Stevenson TG, Goldberg AP. Metabolic inflexibility during submaximal aerobic exercise is associated with glucose intolerance in obese older adults. *Obesity (Silver Spring).* 2014;22(2):451–457.
22. Irving BA, Lanza IR, Henderson GC, Rao RR, Spiegelman BM, Nair KS. Combined training enhances skeletal muscle mitochondrial oxidative capacity independent of age. *J Clin Endocrinol Metab.* 2015;100(4):1654–1663.
23. Robinson MM, Dasari S, Konopka AR, Johnson ML, Manjunatha S, Esponda RR, Carter RE, Lanza IR, Nair KS. Enhanced protein translation underlies improved metabolic and physical adaptations to different exercise training modes in young and old humans. *Cell Metab.* 2017;25(3):581–592.
24. Gavin TP, Kraus RM, Carrithers JA, Garry JP, Hickner RC. Aging and the skeletal muscle angiogenic response to exercise in women. *J Gerontol A Biol Sci Med Sci.* 2015;70(10):1189–1197.
25. Hamilton MT. The role of skeletal muscle contractile duration throughout the whole day: reducing sedentary time and promoting universal physical activity in all people. *J Physiol.* 2017. doi: 10.1113/JP273284. (Epub ahead of print).
26. Hamilton MT, Healy GN, Dunstan DW, Zderic TW, Owen N. Too little exercise and too much sitting: inactivity physiology and the need for new recommendations on sedentary behavior. *Curr Cardiovasc Risk Rep.* 2008;2(4):292–298.
27. Levine JA. Sick of sitting. *Diabetologia.* 2015;58:1751–1758.
28. Spittaels H, Van Cauwenberghe E, Verbestel V, De Meester F, Van Dyck D, Verloigne M, Haerens L, Deforche B, Cardon G, De Bourdeaudhuij I. Objectively measured sedentary time and physical activity time across the lifespan: a cross-sectional study in four age groups. *Int J Behav Nutr Phys Act.* 2012;9:149.
29. Schuna Jr JM, Johnson WD, Tudor-Locke C. Adult self-reported and objectively monitored physical activity and sedentary behavior: NHANES 2005-2006. *Int J Behav Nutr Phys Act.* 2013;10:126.
30. Loprinzi PD. Accelerometer-determined sedentary and physical activity estimates among older adults with diabetes: considerations by demographic and comorbidity characteristics. *J Aging Phys Act.* 2013;22:432–440.

31. Finni T, Haakana P, Pesola AJ, Pullinen T. Exercise for fitness does not decrease the muscular inactivity time during normal daily life. *Scand J Med Sci Sports.* 2014;24(1): 211–219.
32. Kozey-Keadle S, Staudenmayer J, Libertine A, Mavilia M, Lyden K, Braun B, Freedson P. Changes in sedentary time and physical activity in response to an exercise training and/or lifestyle intervention. *J Phys Act Health.* 2014;11(7):1324–1333.
33. Bey L, Areiqat E, Sano A, Hamilton MT. Reduced lipoprotein lipase activity in postural skeletal muscle during aging. *J Appl Physiol.* 2001;91:687–692.
34. Howard BJ, Fraser SF, Sethi P, Cerin E, Hamilton MT, Owen N, Dunstan DW, Kingwell BA. Impact on hemostatic parameters of interrupting sitting with intermittent activity. *Med Sci Sports Exerc.* 2013;45(7):1285–1291.
35. Dunstan DW, Kingwell BA, Larsen R, Healy GN, Cerin E, Hamilton MT, Shaw JE, Bertovic DA, Zimmet PZ, Salmon J, Owen N. Breaking up prolonged sitting reduces postprandial glucose and insulin responses. *Diabetes Care.* 2012;35(5):976–983.
36. Stephens BR, Granados K, Zderic TW, Hamilton MT, Braun B. Effects of 1 day of inactivity on insulin action in healthy men and women: interaction with energy intake. *Metabolism.* 2011;60(7):941–949.
37. Hamilton MT, Areiqat E, Hamilton DG, Bey L. Plasma triglyceride metabolism in humans and rats during aging and physical inactivity. *Int J Sport Nutr Exerc Metab.* 2001;11(suppl):S97–S104.
38. DeFronzo RA, Ferrannini E, Sato Y, Felig P, Wahren J. Synergistic interaction between exercise and insulin on peripheral glucose uptake. *J Clin Invest.* 1981;68 (6):1468–1474.
39. Dunaif A. Insulin resistance and the polycystic ovary syndrome: mechanism and implications for pathogenesis. *Endocr Rev.* 1997;18:774–800.
40. Ardigo D, Numeroso F, Valtuena S, Franzini L, Piatti PM, Monti L, Delsignore R, Reaven GM, Zavaroni I. Hyperinsulinemia predicts hepatic fat content in healthy individuals with normal transaminase concentrations. *Metabolism.* 2005;54: 1566–1570.
41. Reaven GM, Lithell H, Landsberg L. Hypertension and associated metabolic abnormalities: the role of insulin resistance and the sympathoadrenal system. *N Engl J Med.* 1996;334:374–381.
42. Reaven GM. The kidney: an unwilling accomplice in syndrome X. *Am J Kidney Dis.* 1997;30:928–931.
43. Reaven GM. Compensatory hyperinsulinemia and the development of an atherogenic lipoprotein profile: the price paid to maintain glucose homeostasis in insulin-resistant individuals. *Endocrinol Metab Clin North Am.* 2005;34:49–62.
44. Corkey BE. Banting lecture 2011: hyperinsulinemia: cause or consequence? *Diabetes.* 2012;61(1):4–13.
45. Schiaffino S, Reggiani C. Fiber types in mammalian skeletal muscles. *Physiol Rev.* 2011;91(4):1447–1531.
46. Laughlin MH, Schrage WG. Effects of muscle contraction on skeletal muscle blood flow: when is there a muscle pump? *Med Sci Sports Exerc.* 1999;31(7):1027–1035.
47. Dyck DJ, Bonen A. Muscle contraction increases palmitate esterification and oxidation and triacylglycerol oxidation. *Am J Physiol.* 1998;275(5 Pt 1):E888–E896.
48. Goldberg IJ, Le NA, Ginsberg HN, Krauss RM, Lindgren FT. Lipoprotein metabolism during acute inhibition of lipoprotein lipase in the cynomolgus monkey. *J Clin Invest.* 1988;81(2):561–568.
49. Ziouzenkova O, Perrey S, Asatryan L, Hwang J, MacNaul KL, Moller DE, Rader DJ, Sevanian A, Zechner R, Hoefler G, Plutzky J. Lipolysis of triglyceride-rich lipoproteins generates PPAR ligands: evidence for an antiinflammatory role for lipoprotein lipase. *Proc Natl Acad Sci USA.* 2003;100(5):2730–2735.

50. Morino K, Petersen KF, Sono S, Choi CS, Samuel VT, Lin A, Gallo A, Zhao H, Kashiwagi A, Goldberg IJ, Wang H, Eckel RH, Maegawa H, Shulman GI. Regulation of mitochondrial biogenesis by lipoprotein lipase in muscle of insulin-resistant offspring of parents with type 2 diabetes. *Diabetes*. 2012;61(4):877–887.

FURTHER READING

Hamilton MT, Owen N, Sedentary behavior and inactivity physiology. In: Bouchard C, Blair SN, Haskell WL, eds. *Physical Activity and Health*. 2. Human Kinetics; 2011,2: 53–68.

CHAPTER SIX

Dwarf Mice and Aging

Michal M. Masternak*,†, Justin Darcy‡, Berta Victoria*, Andrzej Bartke‡,1
* University of Central Florida, Orlando, FL, United States
† The Greater Poland Cancer Centre, Poznan, Poland
‡ Southern Illinois University School of Medicine, Springfield, IL, United States
1 Corresponding author. E-mail address: abartke@siumed.edu

Contents

Abstract

Dwarf mice have been studied for many decades, however, the focus of these studies shifted in 1996 when it was shown by Brown-Borg and her coworkers that Ames dwarf (*Prop1^{df}*) mice are exceptionally long-lived. Since then, Snell dwarf (*Pit1^{dw}*) and growth hormone receptor knockout (GHR-KO, a.k.a. Laron dwarf) mice were also shown to be exceptionally long-lived, presumably due to their growth hormone (GH)-deficiency or -resistance, respectively. What is of equal importance in these dwarf mice is their extended health span, that is, these animals have a longer period of life lived free of frailty and age-related diseases. This review article focuses on recent studies conducted in these dwarf mice, which concerned brown and white adipose tissue biology, microRNA (miRNA) profiling, as well as early-life dietary and hormonal interventions. Results of these studies identify novel mechanisms linking reduced GH action with extensions of both life span and health span.

Progress in Molecular Biology and Translational Science, Volume 155
ISSN 1877-1173
http://dx.doi.org/10.1016/bs.pmbts.2017.12.002

1. INTRODUCTION

Mice with hereditary dwarfism have been used in biomedical research for nearly 90 years. In 1929, George Snell discovered a recessive mutation that in a homozygous state caused drastic retardation of postnatal growth and a diminutive adult phenotype,[1] which was later shown to be related to defective development of the anterior pituitary and absence of growth hormone (GH) producing cells.[2,3] Because of the lack of GH action in these dwarf (dw) mice (now known as Snell dwarf mice), they were used extensively to study the physiological actions of GH and consequences of GH deficiency (GHD).[4] Snell dwarf mice are also thyroid-stimulating hormone (TSH) deficient,[3] and therefore are used to study the effects of hypothyroidism. Approximately 30 years later, Robert Schaible found a recessive mutant with an identical phenotype, determined that this mutation was not allelic to Snell dwarfism and, in fact, was located on a different chromosome. This new mutant was named Ames dwarf (df).[5] A few years later, both Snell and Ames dwarf mice were shown not only to be GH- and TSH-deficient, but also prolactin (PRL)-deficient,[6] and thus uniquely suitable for the study of the actions of PRL.[7] Elegant cloning studies in the Rosenfield laboratory identified deficits in the genes mutated in the Snell dwarf and the Ames dwarf mouse [*Pituitary Factor 1* (*Pit1*) and *Prophet of Pituitary Factory 1* (*Prop1*), respectively], which confirmed deficiency of GH, PRL, and TSH in these mutants.[3,8]

The first indication that these mutations influence aging came from studies conducted in the early 1970s. Curiously, one laboratory reported that Snell dwarf mice are exceptionally short-lived due to immune deficits and susceptibility to infections,[9] while another laboratory used these animals to study the impact of delayed aging on osteoarthritis because they lived much longer than their normal siblings.[10,11] We now believe that the short life span of Snell dwarf mice studied by Fabris *et al.*[9] must have been due to husbandry practices or pathogens at his institution, as other investigators did not encounter similar symptoms.[12] In 1996, Brown-Borg *et al.* reported that Ames dwarf mice in their colonies outlived their normal siblings by approximately 1 year, or roughly 50%.[13] The extension of longevity was somewhat greater in females, but average, as well as maximal longevity were significantly increased in both sexes. Flurkey *et al.* demonstrated that Snell dwarf mice also live significantly longer than their normal siblings,[14] and provided important evidence that this is associated with delayed aging of the immune system and collagen fibers.[14] These findings, along with subsequent studies of other characteristics that normally change with age, provided evidence that extended longevity of mice with genetic dwarfism is accompanied by, and almost certain due to slower and/or delayed aging.[15]

In 2000, Coschigano *et al.* reported that mice with targeted deletion of the GH receptor gene (resulting in GH-resistance) have markedly extended longevity,[16] an observation that was confirmed and extended in further studies involving GH-receptor deleted (GHR-KO) mice on two different genetic backgrounds.[17] This work firmly established that reducing GH signaling can be sufficient to increase life span of a mammal, and supported the suggestion that the extended longevity of Ames and Snell dwarf mice is due primarily to GHD. The link of reduced GH signaling to slow aging is further supported by the demonstration that excessive GH lends to accelerated aging and truncated life span in mice and possibly humans.[18] Studies conducted in the Ruvkun laboratory provided evidence that Daf-2, the key longevity gene in the microscopic worm, *Caenorhabditis elegans*, is homologous to mammalian insulin-like growth factor 1 (IGF-1) receptor and insulin receptor genes.[19] As IGF-1 is an important mediator of GH actions in vertebrates, and insulin secretion and action are modified by GH, it quickly became apparent that homologous signaling pathways influence aging in organisms ranging from unicellular yeast, through worms and insects, to mammals, including humans.[20–23] The evolutionary conservation of the mechanism that control aging is really astounding when considered against the background of differences in the life cycles and physiology of the organisms involved. Following the discovery of the role of GH in mammalian aging, work in a number of laboratories was directed at identifying the underlying mechanisms involved. The picture emerging from these studies is a network of interacting mechanisms that include changes in mitochondrial function, redox mechanisms, mTOR signaling, inflammation, glucose homeostasis, energy metabolism, and stress resistance. Much of this work, along with the controversies surrounding the role of GH in human aging, have been recently reviewed and interested readers are referred to these articles.[15,20,24,25] In this brief review, we will not attempt to present or discuss all of the work related to the delayed aging and extended longevity of dwarf mice, but instead, we will describe findings from a few recent studies that dealt with these intriguing animals and provided new insights.

2. ADIPOSE TISSUE

Adipose tissue was initially believed to be a metabolically inert organ, with its main function being the storage of energy in the form of lipids, as well as act as insulation. However, extensive studies over the past several decades indicate that adipose tissue is a major endocrine organ that regulates whole-body metabolism through interactions with different insulin target organs. Consisting of loose connective tissue mostly composed of adipocytes, adipose tissue also contains preadipocytes, mesenchymal stem cells, fibroblasts, vascular endothelial cells, adipose tissue macrophages, and other types of immune cells, including regulatory T cells (Tregs).[26] This complex composition appears to modulate different endocrine activities of adipose tissue by either promoting insulin sensitivity and healthy metabolism, or metabolic syndrome, insulin resistance, diabetes, and systemic inflammation.[26–30] The major (and the most studied) endocrine function of adipose tissue includes the production and release of different hormones (referred to as adipokines), including adiponectin, estrogen, leptin, resistin, and a variety of cytokines. Production of these cytokines and their release into circulation affects the function of other hormones by interacting with their receptors.

3. QUALITY VERSUS QUANTITY: DIVERGENT ROLE OF FAT IN LONG-LIVING ANIMALS

There are two major types of adipose tissue: white adipose tissue (WAT) and brown adipose tissue (BAT). The main function of BAT is to generate body heat through nonshivering thermogenesis (NST); whereas WAT represents the main tissue that stores energy and regulates metabolic functions. There is a well-established correlation between the quantity of WAT and its endocrine function, that is, "quality." High accumulation of WAT in obese individuals, which relates to an increased body mass index (BMI) represents increased health risks including the development of insulin resistance, metabolic syndrome, diabetes, cardiovascular disease (CVD), and cancer. Importantly, higher fat accumulation also causes an increase in the size of adipocytes, which might contribute to the increased inflammatory status and release of harmful proinflammatory cytokines. Opposite to this, long-living dwarf mice, and animals subjected to calorie restriction (CR), have smaller adipocytes.[31] However, the function of WAT does not only depends on the quantity and the size of adipocytes, but also on its location within the body. Importantly, different adipose tissue depots have different roles in whole-body metabolism, which relates to the release of different panels of cytokines and adipokines. Subcutaneous fat is generally considered to be "good" WAT, and an increased accumulation of this adipose depot usually does not cause any metabolic complications.[32] More importantly, some studies show that performing liposuction of subcutaneous fat does not provide any improvement of metabolic status beside cosmetic benefits. On the other hand, visceral fat removal in humans appears to improve metabolic status, whereas the same surgical procedures in rodents produces significant improvements of insulin sensitivity with a concomitant increase in life span.[33,34] These and other studies show that mainly visceral fat represents a risk of metabolic complications. However, our studies with long-lived mice characterized by GH-resistance or -deficiency showed that in some cases, the function of different adipose depots is altered, and that visceral fat in these animals stimulates improved metabolism (resembling the function of subcutaneous adipose tissue).[35–37] For example, long-lived GHR-KO and Ames dwarf mice are metabolically healthy with improved insulin sensitivity and glucose tolerance, yet they have a tendency to become obese as they age, showing an increase of both subcutaneous and visceral adipose depots.[35–41] Hyperinsulinemic-euglycemic clamp studies in Ames dwarf mice showed that to maintain equal glycemia between normal and dwarf mice, the dwarf animals had to be infused with double the dosage of glucose.[40] This test indicted improvement of whole-body insulin sensitivity with much more dramatic hepatic responses to glucose infusion in dwarf mice when compared to their normal littermates, indicating the liver as a major organ in promoting insulin sensitivity.[40] However, the analysis of glucose uptake during the clamp study indicated a significant increase in glucose clearance by dwarf WAT compared to the same adipose depot in their normal littermates.[40] Importantly, further studies indicated an increase in the major insulin signaling and metabolic pathways in epididymal adipose tissue of long-living dwarf and GHR-KO mice, which included increased levels of insulin receptor, insulin receptor substrate 1, PGC1a, PPAR-γ, and PPAR-α.[35,36,42] Visceral fat from these GH-deficient or -resistant mice is also characterized by decreased levels of inflammatory IL-6 and TNF-α, indicating that regardless of the increased amount of visceral fat in these animals, there is no concomitant increase of inflammation as confirmed by analysis of serum levels for IL-6 and TNF-α.[35,36] Overall, these studies showed that visceral fat from long-lived dwarf and GHR-KO mice is significantly different than the same adipose depot from their normal littermates. However, the important question is how different is the endocrine function and regulation of metabolism by these depots in dwarf and GHR-KO mice. Our studies of visceral fat removal (VFR) in normal mice confirmed previous findings that removing visceral fat improves insulin sensitivity and glucose tolerance,[35,36] while the same surgical intervention in GHR-KO and dwarf mice caused a decrease in insulin sensitivity and glucose tolerance, or no improvements in these parameters of glucose homeostasis.[35,36] This study indicated that visceral fat, and especially the quantity of this fat in normal animals, has a negative impact on metabolism, however, it is a genetic modification due to GH signaling suppression, which alters the function of visceral WAT in dwarf and GHR-KO mice. It could be argued that either (1) WAT from dwarf and GHR-KO mice produces critical adipokines, such as adiponectin, that promote insulin sensitivity, or (2) removing large quantities of WAT outbalance metabolic status in these healthy obese animals causing metabolic shock that subsequently results in the disruption of glucose clearance and impairment of health status. However, another study performed by our research team showed that transplanting visceral fat from GHR-KO mice to their normal littermates causes improvement of glucose tolerance and insulin sensitivity in the recipient mice.[43] More importantly, this study showed that besides systemic improvements of insulin sensitivity, implanted WAT from GHR-KO mice caused improvement of the insulin signaling pathway in the liver of these animals.[43] This demonstrates that implanted healthy visceral WAT not only adds to glucose clearance, but also stimulates hepatic insulin sensitivity through its hormonal activity.[43] Our study with a high fat diet regimen showed that similar experimentation with fat transplant between dwarf and normal littermates provided protection against the negative effects of high fat diet.[44] Following these physiological observations, in depth transcriptome studies of visceral fat exhibited a divergent shift in the gene expression of BAT and WAT from these long-lived mice.[37] As expected, there were major difference between BAT and WAT, however, BAT from GHR-KO mice indicated increased expression levels in genes regulating mitochondria biogenesis and metabolism with opposite regulation of genes expressed mainly by dendritic cells and macrophages, indicating reduced inflammatory status and decreased infiltration by macrophages in BAT when compared to WAT.[37] This study showed that both BAT and WAT transcriptomes are altered due to silencing of GH signaling in GHR-KO mice when compared with their normal littermates.[37] More importantly, comparisons of epididymal, perirenal, and subcutaneous WAT in this study indicated that these different fat pads are characterized by stronger distinction in normal mice than in GHR-KO animals. This suggests that the visceral fat transcriptome (epididymal and perirenal) in GHR-KO mice to some extent resembled the transcriptome of metabolically healthy subcutaneous fat, which can explain beneficial endocrine role of these fat depots on metabolism in GHR-KO animals or in normal mice after transplant.[37]

4. MicroRNA (miRNA) AND HEALTHY AGING

Maintenance of homeostasis at the molecular, cellular, organ, and organismal level depends on the complex network of hormonal and cytokine signals. Among other regulatory inputs, there is a growing interest in small noncoding RNAs as important factors that regulate the transcriptome, and represent a tool for communication between distant organs.[45–49] MicroRNA (miRNA) is known to be involved in aging, metabolism, neurodegenerative diseases, and cancer. Worth mentioning are circulating miRNAs that are encapsulated in exosomes that can travel in circulation to different organs and have hormone-like activity through posttranscriptional regulation of various genes. miRNAs are known to suppress translation, but apparently can also stabilize mRNA, or even stimulate translation. However, the exact mechanisms of these different actions are not well-established. It is also known that circulating miRNAs can be produced by a variety of cells in different tissues, and that the expression levels of different miRNAs are affected by disease and aging.

High throughput profiling of circulating miRNA in young and old Ames dwarf mice and their normal littermates revealed genotype-specific changes of the expression levels of 21 circulating miRNAs during aging, indicating significant genotype by age (GbA) interaction.[45] Interestingly, 17 out of 21 miRNA showed a general trend toward downregulation in circulation during normal aging (including miR-344d-3-5p, miR-376c-5p, miR-136-5p, miR-411-5p, miR-344d-1-5p, miR-410-5p, miR-369-5p, miR-154-5p, miR-540-5p, miR-127-5p, miR-449a-5p, miR-381-5p, miR-344d-3p, and miR-195a-5p); whereas the same miRNAs were maintained at youthful level in long-lived dwarf mice as they aged, or were downregulated during aging in normal mice but overexpressed in dwarf mice (including miR-34c-5p, miR-34b-5p, and miR-344d-2-5p). There was also a group of three miRNAs (miR-5107-5p, miR-146a-5p, and miR-342-5p) that increased with age in normal mice, but stayed unchanged in long-lived dwarf mice.[45]

MiR-592-5p was the only miRNA that did not change in normal mice during aging, and was elevated in dwarf mice.[45] Interestingly, we have shown that some of these miRNA families (including miR-34, miR-146, and miR-449) are similarly regulated by CR and in dwarf mice during healthy aging.[45] There are also several reports indicating differential regulation of these miRNA in different tissues and in circulation.[46,50–52] For example, miR-146a-5p was found to be downregulated in aged CR and dwarf animals when compared to ad libitum-fed normal animals.[45,46,51] miR-146a-5p plays an important role in regulating the innate immune response by regulating Toll-Like Receptors (TLRs) and stimulating the production of some proinflammatory cytokines.[46] We could argue that the expression of this miRNA is important for a proper immune response, however, significant overexpression in aged animals can disrupt proper functioning of TLRs and promote chronic inflammation. Other studies have showed that senescent cells are characterized by overexpression of miR-146a, which could explain increased levels of this miRNA during aging, and could be suggested as a possible mechanism of inflammaging.[46] On the other hand, upregulated in dwarf circulation, miR-34b was also found to be elevated in the livers of C57BL/6 and Keap1-KD mice when subjected to CR, suggesting hepatocytes as a potential source of circulating miR-34b.[46] Moreover, a recent study published by Thomou *et al.* revealed adipose tissue as one of major source of circulating exosomal miRNA that can regulate the expression of different genes in different insulin target organs, including liver.[53] More importantly, select miRNAs showing GbA interaction in dwarf studies are involved in regulating important processes of healthy aging. Our functional enrichment analyses of validated miRNA targets indicated involvement in tumor suppression pathways, antiinflammatory responses, as well as modulation of Wnt, mTOR, AMPK, and insulin signaling pathways.[46] The study by Bates *et al.* showed that another miRNA, miR-27a, is an important posttranscriptional regulator of glutathione metabolism, the urea cycle, and polyamine biosynthesis, which interestingly were observed to be differentially expressed as early as 2 months of age in the liver of dwarf mice compared with their normal littermates.[54] Other groups of miRNAs, including miR-470, miR669b, and miR-681, were shown to be upregulated in dwarf mice. These miRNAs represent putative main regulators of GH/IGF-1 signaling.[55] The regulatory role of these three miRNAs is suggested to modulate phosphorylation of PI3K, AKT, and FoxO3a, which are major signaling proteins involved in longevity. These studies showed the potential importance of the mechanistic role of miRNAs in regulating different pathways related to aging and longevity, with putative hormone-like activity allowing for interactive communications between different organs.[46,53]

5. ENERGY METABOLISM

Recently, energy metabolism has garnered considerable interest due to its possible use to combat the growing obesity and diabetes epidemics. Of note, two output parameters of energy metabolism that can be measured by indirect calorimetry are oxygen consumption (VO_2) and respiratory quotient (RQ). VO_2 is directly related to the metabolic rate of an animal, and RQ is indicative of the macromolecules used as an energy source of an animal (RQ ranges from 0.7 to 1.0, where 0.7 is pure lipid utilization and 1.0 is pure carbohydrate utilization). Contrary to our laboratories initial hypothesis, Ames dwarf and GHR-KO mice have increased VO_2 per gram body weight, and we have also shown that they have reduced RQ.[56]

While this observed increased in VO_2 seems counter to the long-standing "Rate of Living Hypothesis,"[57] it is worth mentioning that this observation falls in line with new (and seemingly more applicable to mammals) hypothesis, the "Uncoupling to Survive Hypothesis".[58–60] According to this hypothesis, uncoupling of the mitochondrial electron transport chain to increase the dissipation of energy as heat is associated with reduced production of the noxious reactive oxygen species and with various metabolic and survival-related benefits (Fig. 1). Evidence that nonshivering thermogenesis via enhanced mitochondrial uncoupling is, indeed, increased in the long-lived dwarf mice is discussed in the next section of this article. As expected, transgenic mice expressing high levels of bovine GH (bGH mice) have the opposite phenotype, with decreased VO_2 per gram body weight and increased RQ. Arguably, the decreased RQ observed in Ames dwarf and GHR-KO mice may be the most important, as an increase in lipid utilization has been observed in many life extending interventions, including CR.[61]

6. SURGICAL REMOVAL OF BROWN ADIPOSE TISSUE IN AMES DWARF MICE

As mentioned earlier, it appears that GH action plays a role in both energy metabolism and adipose tissue function. Therefore, BAT became a logical tissue to examine in dwarf mice. The first indication that BAT is

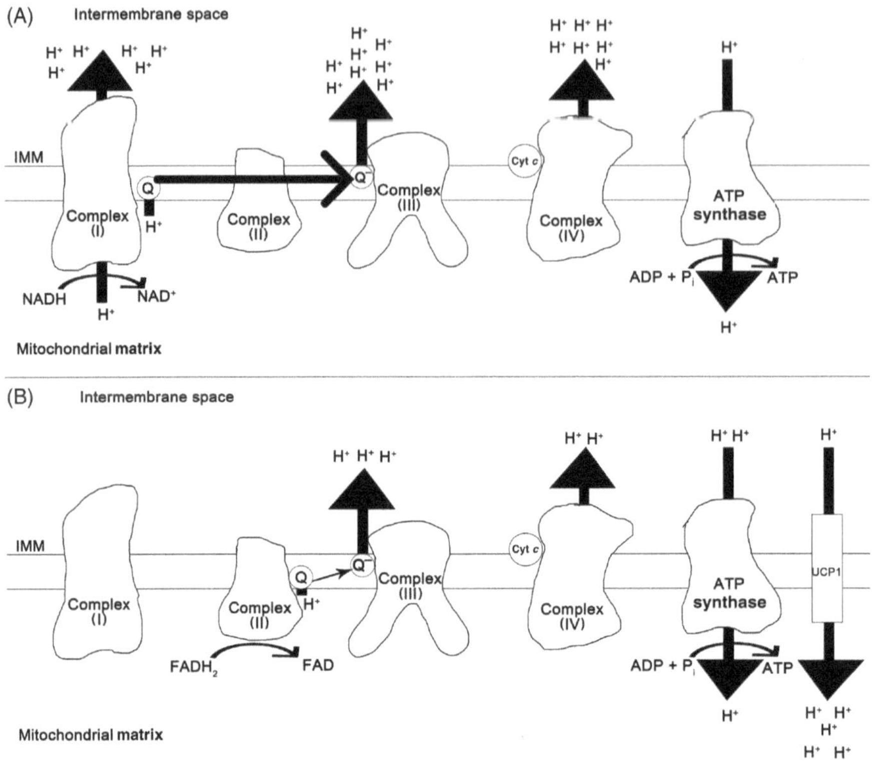

Fig. 1 Uncoupling and reactive oxygen species (ROS) production. (A) Under normal physiological conditions, NADH from the Kreb's cycle is used as a reduced electron carrier at Complex (I) of the electron transport chain (ETC). However, Complex (I) is a major site of electron leak, which results in increased ROS production. Quinone (Q) carries electrons from Complex (I) to Complex (III), however, due to the high inner mitochondrial membrane (IMM) potential, reduced Q stalls at Complex (III), resulting in electron leak and subsequent ROS production. (B) In brown adipose tissue (where uncoupling is extremely high), there is an increase in the molar ratio of $FADH_2$ produced compared to NADH due to an increase in fatty acid β-oxidation. This results in bypassing Complex (I); and therefore, electron leak and ROS production. Moreover, uncoupling protein 1 (UCP1) uncouples the ETC, which results in a lower IMM potential. The decreased membrane potential results in less stalling of Q, and therefore less electron leak at Complex (III).

altered in dwarf mice was observed in GHR-KO mice, which were shown to have increased relative weight of the interscapular BAT depot along with increased expression of the gene and protein levels of UCP-1 in BAT.[62] This increased expression of UCP-1 may in part be due to the increased levels of

FGF-21,[63] which has been shown to stimulate UCP-1 expression.[64] Moreover, this study established that bGH transgenic mice have the opposite phenotype,[62] demonstrating GH signaling is directly involved in BAT function. Moreover, growth hormone releasing hormone knockout (GHRH-KO) mice also have an increase in the relative weight of their interscapular BAT and UCP-1 gene expression.[65]

Recently, our laboratory demonstrated that Ames dwarf mice also have alterations in their BAT compared to their normal littermates including an increase in the relative weight of the interscapular BAT depot, increased expression of genes related to thermogenesis and lipid metabolism, and observable differences in tissue sections stained with hematoxylin and eosin, such as depleted lipid vacuoles.[66] Interestingly, surgical removal of the interscapular BAT depot resulted in opposite physiological responses where Ames dwarf mice lost adiposity and normal mice gained adiposity.[66] We believe this unusual finding in Ames dwarf mice may in part be due to their lower core body temperature.[67] Studies involving concurrent BAT removal and exposure to thermoneutrality (30°C) in Ames dwarf mice and normal mice are of extreme interest to our laboratory and are currently ongoing.

7. EARLY-LIFE INTERVENTIONS

It has been well established that early-life events can alter the overall health of an individual. While this indirectly may lead to alterations in longevity (e.g., a fetus of a diabetic mother is at an increased risk for metabolic dysfunction, which may shorten their life expectancy), it is desirable to evaluate early-life interventions that directly alter longevity in and of themselves. Potential of early-life interventions to extend longevity was first established by Sun *et al.* through the use of litter crowding during the preweaning period, which artificially creates a mild CR. Since then, our laboratory has become increasingly interested in the effects of early-life programming of aging. We have demonstrated that a 6-week treatment period of GH early in life shortens longevity in Ames dwarf mice, while a 6-week treatment period of thyroxine early in life does not (unpublished data and Refs. [68,69]68 and 69). Interestingly, several of the key parameters affected by early-life GH treatment (e.g., glucose and energy metabolism) remain altered even 1 year after completion of GH treatment.[70]

8. REVIEW SUMMARY

Dwarf mice are unique in that they have extraordinarily extended life span and health span. Our previous studies of these animals were focused on their extreme insulin sensitivity as a means for extended longevity. While we believe insulin signaling to be critical to their longevity, recent studies involving adipose tissue and microRNAs, as well as studies focused on energy metabolism, have provided novel insights into the possible mechanisms underlying the longevity and health span of dwarf mice. In this review, we highlighted some of these recent findings, as well as provided insight into areas we believe need to be further examined, or are currently being explored.

ACKNOWLEDGMENTS

This work was supported by the National Institutes of Health (NIH) grant AG051860. We would like to acknowledge Lisa Hensley for editorial assistance, as well as contributions and support from past and present laboratory members and from our collaborators. We sincerely apologize to those whose work pertinent to this topic was not mentioned either due to space limitations or inadvertent omission. This work is solely the view of the authors, and does not reflect those of the NIH.

REFERENCES

1. Snell GD. Dwarf, a new mendelian recessive character of the house mouse. *Proc Natl Acad Sci USA*. 1929;15:733–734.
2. Carsner RL, Rennels EG. Primary site of gene action in anterior pituitary dwarf mice. *Science*. 1960;131(3403):829.
3. Li S, Crenshaw 3rd EB, Rawson EJ, Simmons DM, Swanson LW, Rosenfeld MG. Dwarf locus mutants lacking three pituitary cell types result from mutations in the POU-domain gene pit-1. *Nature*. 1990;347(6293):528–533.
4. Bartke A. Genetic models in the study of anterior pituitary hormones. In: Shire JGM, ed. In: *Genetic Variation in Hormone Systems. Boca Raton: CRC Press; 1979:113–126.*
5. Schaible R, Gowen JW. A new dwarf mouse. *Genetics*. 1961;46:896.
6. Bartke A. Influence of luteotrophin on fertility of dwarf mice. *J Reprod Fertil.* 1965;10:93–103.
7. Bartke A. Prolactin-deficient mice. In: Alexander NJ, ed. In: *Animal Models for Research on Contraception and Fertility. Hagerstown: Harper & Row; 1979:360–365.*
8. Andersen B, Pearse 2nd RV, Jenne K, et al. The Ames dwarf gene is required for Pit-1 gene activation. *Dev Biol.* 1995;172(2):495–503.
9. Fabris N, Pierpaoli W, Sorkin E. Hormones and the immunological capacity III: the immuno-deficient disease of the hypopituitary Snell-Bagg dwarf mouse. *Clin Exp Immunol.* 1971;9:109–225.
10. Silberberg R. Articular aging and osteoarthrosis in dwarf mice. *Pathol Microbiol.* 1972;38:417–430.

11. Silberberg R. Vertebral aging in hypopituitary dwarf mice. *Gerontologia.* 1973;19(5):281–294.
12. Shire JG. Growth hormone and premature ageing. *Nature.* 1973;245(5422):215–216.
13. Brown-Borg HM, Borg KE, Meliska CJ, Bartke A. Dwarf mice and the ageing process. *Nature.* 1996;384(6604):33.
14. Flurkey K, Papaconstantinou J, Miller RA, Harrison DE. Lifespan extension and delayed immune and collagen aging in mutant mice with defects in growth hormone production. *Proc Natl Acad Sci USA.* 2001;98(12):6736–6741.
15. Bartke A. Single-gene mutations and healthy ageing in mammals. *Philos Trans R Soc Lond B Biol Sci.* 2011;366(1561):28–34.
16. Coschigano KT, Clemmons D, Bellush LL, Kopchick JJ. Assessment of growth parameters and lifespan of GHR/BP gene-disrupted mice. *Endocrinology.* 2000;141:2608–2613.
17. Coschigano KT, Holland AN, Riders ME, List EO, Flyvbjerg A, Kopchick JJ. Deletion, but not antagonism, of the mouse growth hormone receptor results in severely decreased body weights, insulin, and insulin-like growth factor I levels and increased life span. *Endocrinology.* 2003;144(9):3799–3810.
18. Bartke A. Can growth hormone (GH) accelerate aging? Evidence from GH-transgenic mice. *Neuroendocrinology.* 2003;78:210–216.
19. Kimura KD, Tissenbaum HA, Liu Y, Ruvkun G. daf-2, an insulin receptor-like gene that regulates longevity and diapause in *Caenorhabditis elegans. Science.* 1997;277(5328):942–946.
20. Bartke A, Sun LY, Longo V. Somatotropic signaling: trade-offs between growth, reproductive development, and longevity. *Physiol Rev.* 2013;93(2):571–598.
21. Kenyon C. The plasticity of aging: insights from long-lived mutants. *Cell.* 2005;120(4):449–460.
22. Longo VD, Finch CE. Evolutionary medicine: from dwarf model systems to healthy centenarians? *Science.* 2003;299(5611):1342–1346.
23. Tatar M, Bartke A, Antebi A. The endocrine regulation of aging by insulin-like signals. *Science.* 2003;299(5611):1346–1351.
24. Brown-Borg HM. The somatotropic axis and longevity in mice. *Am J Physiol Endocrinol Metab.* 2015;309(6):E503–E510.
25. Bartke A, List EO, Kopchick JJ. The somatotropic axis and aging: benefits of endocrine defects. *Growth Horm IGF Res.* 2016;27:41–45.
26. Martyniak K, Masternak MM. Changes in adipose tissue cellular composition during obesity and aging as a cause of metabolic dysregulation. *Exp Gerontol.* 2016;94:59–63.
27. Cipolletta D, Cohen P, Spiegelman BM, Benoist C, Mathis D. Appearance and disappearance of the mRNA signature characteristic of Treg cells in visceral adipose tissue: age, diet, and PPARgamma effects. *Proc Natl Acad Sci USA.* 2015;112(2):482–487.
28. Cipolletta D, Kolodin D, Benoist C, Mathis D. Tissular T(regs): a unique population of adipose-tissue-resident Foxp3+CD4+ T cells that impacts organismal metabolism. *Semin Immunol.* 2011;23(6):431–437.
29. Feuerer M, Herrero L, Cipolletta D, et al. Lean, but not obese, fat is enriched for a unique population of regulatory T cells that affect metabolic parameters. *Nat Med.* 2009;15(8):930–939.
30. Kolodin D, van Panhuys N, Li C, et al. Antigen- and cytokine-driven accumulation of regulatory T cells in visceral adipose tissue of lean mice. *Cell Metab.* 2015;21(4):543–557.
31. Wiesenborn DS, Menon V, Zhi X, et al. The effect of calorie restriction on insulin signaling in skeletal muscle and adipose tissue of Ames dwarf mice. *Aging.* 2014;6(10):900–912.
32. Tran TT, Yamamoto Y, Gesta S, Kahn CR. Beneficial effects of subcutaneous fat transplantation on metabolism. *Cell Metab.* 2008;7(5):410–420.

33. Gabriely I, Ma XH, Yang XM, et al. Removal of visceral fat prevents insulin resistance and glucose intolerance of aging: an adipokine-mediated process? *Diabetes.* 2002;51 (10):2951–2958.
34. Muzumdar R, Allison DB, Huffman DM, et al. Visceral adipose tissue modulates mammalian longevity. *Aging Cell.* 2008;7(3):438–440.
35. Masternak MM, Bartke A, Wang F, et al. Metabolic effects of intra-abdominal fat in GHRKO mice. *Aging Cell.* 2012;11(1):73–81.
36. Menon V, Zhi X, Hossain T, et al. The contribution of visceral fat to improved insulin signaling in Ames dwarf mice. *Aging Cell.* 2014;13:497–506.
37. Stout MB, Swindell WR, Zhi X, et al. Transcriptome profiling reveals divergent expression shifts in brown and white adipose tissue from long-lived GHRKO mice. *Oncotarget.* 2015;6(29):26702–26715.
38. Bonkowski MS, Pamenter RW, Rocha JS, Masternak MM, Panici JA, Bartke A. Long-lived growth hormone receptor knockout mice show a delay in age-related changes of body composition and bone characteristics. *J Gerontol A Biol Sci Med Sci.* 2006;61 (6):562–567.
39. Masternak MM, Panici JA, Bonkowski MS, Hughes LF, Bartke A. Insulin sensitivity as a key mediator of growth hormone actions on longevity. *J Gerontol A Biol Sci Med Sci.* 2009;64(5):516–521.
40. Wiesenborn DS, Ayala JE, King E, Masternak MM. Insulin sensitivity in long-living Ames dwarf mice. *Age.* 2014;36(5):9709.
41. Berryman DE, List EO, Palmer AJ, et al. Two-year body composition analyses of long-lived GHR null mice. *J Gerontol A Biol Sci Med Sci.* 2010;65(1):31–40.
42. Bonkowski MS, Dominici FP, Arum O, et al. Disruption of growth hormone receptor prevents calorie restriction from improving insulin action and longevity. *PLoS One.* 2009;4(2):e4567.
43. Bennis MT, Schneider A, Victoria B, et al. The role of transplanted visceral fat from the long-lived growth hormone receptor knockout mice on insulin signaling. *Geroscience.* 2017;39(1):51–59.
44. Hill CM, Fang Y, Miquet JG, Sun LY, Masternak MM, Bartke A. Long-lived hypopituitary Ames dwarf mice are resistant to the detrimental effects of high-fat diet on metabolic function and energy expenditure. *Aging Cell.* 2016;15(3):509–521.
45. Victoria B, Dhahbi JM, Nunez Lopez YO, et al. Circulating microRNA signature of genotype-by-age interactions in the long-lived Ames dwarf mouse. *Aging Cell.* 2015; 14(6):1055–1066.
46. Victoria B, Nunez Lopez YO, Masternak MM. MicroRNAs and the metabolic hallmarks of aging. *Mol Cell Endocrinol.* 2017;455:131–147.
47. Ruvkun G. The perfect storm of tiny RNAs. *Nat Med.* 2008;14(10):1041–1045.
48. Breving K, Esquela-Kerscher A. The complexities of microRNA regulation: mirandering around the rules. *Int J Biochem Cell Biol.* 2010;42(8):1316–1329.
49. Borralho PM, Rodrigues CMP, Steer CJ. Mitochondrial microRNAs and their potential role in cell function. *Curr Pathobiol Rep.* 2014;2(3):123–132.
50. Dhahbi JM. Circulating small noncoding RNAs as biomarkers of aging. *Ageing Res Rev.* 2014;17:86–98.
51. Dhahbi JM, Spindler SR, Atamna H, et al. Deep sequencing identifies circulating mouse miRNAs that are functionally implicated in manifestations of aging and responsive to calorie restriction. *Aging.* 2013;5(2):130–141.
52. Victoria B, Dhahbi JM, Nunez Lopez YO, et al. Circulating microRNA signature of genotype-by-age interactions in the long-lived Ames dwarf mouse. *Aging Cell.* 2015;14 (6):1055–1066.
53. Thomou T, Mori MA, Dreyfuss JM, et al. Adipose-derived circulating miRNAs regulate gene expression in other tissues. *Nature.* 2017;545(7653):252.

54. Bates DJ, Li N, Liang R, et al. MicroRNA regulation in Ames dwarf mouse liver may contribute to delayed aging. *Aging Cell.* 2010;9(1):1–18.
55. Liang R, Khanna A, Muthusamy S, et al. Post-transcriptional regulation of IGF1R by key microRNAs in long-lived mutant mice. *Aging Cell.* 2011;10(6):1080–1088.
56. Westbrook R, Bonkowski MS, Strader AD, Bartke A. Alterations in oxygen consumption, respiratory quotient, and heat production in long-lived GHRKO and Ames dwarf mice, and short-lived bGH transgenic mice. *J Gerontol A Biol Sci Med Sci.* 2009;64(4):443–451.
57. Pearl R. *The Rate of Living.* London, UK: University of London Press; 1928.
58. Brand MD. Uncoupling to survive? The role of mitochondrial inefficiency in ageing. *Exp Gerontol.* 2000;35(6–7):811–820.
59. Speakman JR, Talbot DA, Selman C, et al. Uncoupled and surviving: individual mice with high metabolism have greater mitochondrial uncoupling and live longer. *Aging Cell.* 2004;3(3):87–95.
60. Rose G, Crocco P, De Rango F, Montesanto A, Passarino G. Further support to the uncoupling-to-survive theory: the genetic variation of human UCP genes is associated with longevity. *PLoS One.* 2011;6(12):e29650.
61. Bruss MD, Khambatta CF, Ruby MA, Aggarwal I, Hellerstein MK. Calorie restriction increases fatty acid synthesis and whole body fat oxidation rates. *Am J Physiol Endocrinol Metab.* 2010;298(1):E108–E116.
62. Li Y, Knapp JR, Kopchick JJ. Enlargement of interscapular brown adipose tissue in growth hormone antagonist transgenic and in growth hormone receptor gene-disrupted dwarf mice. *Exp Biol Med.* 2003;228(2):207–215.
63. Brooks NE, Hjortebjerg R, Henry BE, List EO, Kopchick JJ, Berryman DE. Fibroblast growth factor 21, fibroblast growth factor receptor 1, and beta-Klotho expression in bovine growth hormone transgenic and growth hormone receptor knockout mice. *Growth Horm IGF Res.* 2016;30–31:22–30.
64. Fisher FM, Maratos-Flier E. Understanding the physiology of FGF21. *Annu Rev Physiol.* 2016;78:223–241.
65. Leone S, Chiavaroli A, Shohreh R, et al. Increased locomotor and thermogenic activity in mice with targeted ablation of the GHRH gene. *Growth Horm IGF Res.* 2015;25(2):80–84.
66. Darcy J, McFadden S, Fang Y, et al. Brown adipose tissue function is enhanced in long-lived male Ames dwarf mice. *Endocrinology.* 2016;157(12):4744–4753.
67. Hunter WS, Croson WB, Bartke A, Gentry MV, Meliska CJ. Low body temperature in long-lived Ames dwarf mice at rest and during stress. *Physiol Behav.* 1999;67(3):433–437.
68. Panici JA, Harper JM, Miller RA, Bartke A, Spong A, Masternak MM. Early life growth hormone treatment shortens longevity and decreases cellular stress resistance in long-lived mutant mice. *FASEB J.* 2010;24(12):5073–5079.
69. Darcy J, Fang Y, Hill CM, McFadden S, Sun LY, Bartke A. Original research: metabolic alterations from early life thyroxine replacement therapy in male Ames dwarf mice are transient. *Exp Biol Med.* 2016;241(16):1764–1771.
70. Sadagurski M, Landeryou T, Cady G, et al. Growth hormone modulates hypothalamic inflammation in long-lived pituitary dwarf mice. *Aging Cell.* 2015;14(6):1045–1054.

CHAPTER SEVEN

Cellular Metabolism and Aging

Andre Catic[1]
Huffington Center on Aging, Stem Cells and Regenerative Medicine Center, Dan L. Duncan Comprehensive Cancer Center, Baylor College of Medicine, Houston, TX, United States
[1] Corresponding author. E-mail address: catic@bcm.edu

Contents

Abstract

Metabolic changes are hallmarks of aging and genetic and pharmacologic alterations of relevant pathways can extend life span. In this review, we will outline how cellular biochemistry and energy homeostasis change during aging. We will highlight protein quality control, mitochondria, epigenetics, nutrient-sensing pathways, as well as the interplay between these systems with respect to their impact on cellular health.

Progress in Molecular Biology and Translational Science, Volume 155
ISSN 1877-1173
http://dx.doi.org/10.1016/bs.pmbts.2017.12.003

1. INTRODUCTION

"So Nature deals with us, and takes away
our playthings one by one..."
from "Nature" by Henry Wadsworth Longfellow

Virtually every aspect of cell biology is affected by aging. Cellular aging is the result of physical and chemical assaults, of molecular exhaustion, and of physiological attempts to compensate for damage. Examples include toxic adducts to macromolecules, attrition of chromosomes, DNA damage, and impaired organelles.[1,2] Current pharmacological attempts to attenuate aging aim at stimulating cellular repair or at eliminating dysfunctional cells.

Evolutionarily conserved pathways with their built-in redundancies and overlaps appear more resilient to damage. In contrast, phylogenetically newer functions are more vulnerable and developmentally advanced cells seem at a disadvantage when it comes to aging. This vulnerability becomes evident in the functional decline of specialized cells of the adaptive immune system or of cortical neurons, which are more sensitive to age-induced damage than their primal and more resilient counterparts, such as myeloid or mesenchymal cells.

Whereas cells of different types may age at different rates, consistent metabolic changes are observed across cell types.[3,4] In this review, we will discuss key metabolic pathways with relevance for cellular aging and outline how they are connected. Though the effects of aging and metabolism are intertwined, their relationship does not follow a scalable action-and-reaction pattern. Instead, low levels of metabolic impairment can diminish the effects of aging through stimulation of repair systems, a concept known as hormesis, while severe metabolic injury accelerates aging.

2. MITOCHONDRIA AND AGING

Mitochondria are the remnants of proteobacteria that became endosymbionts 1–2 billion years ago. Mitochondria still retain a small genome that encodes 37 genes in animals, but is larger in simpler eukaryotes. The integration of mitochondrial genes into the host's chromosomes comes with advantages as their expression, replication, and repair are under control by the more sophisticated nuclear machinery of eukaryotes. Mitochondria did not abandon their entire genome, though, and retained a small plasmid that encodes some subunits of the respiratory chain complex. The topology of these proteins, being inserted into the inner mitochondrial membrane, is the likely reason why they are translated inside mitochondria, whereas nuclear encoded proteins are translated in the cytosol and may require import through the outer and inner mitochondrial membranes.[5]

In contrast to their neglect of DNA housekeeping, mitochondria have become the central hub for metabolism of the host cells. They not only produce ATP, but also are essential for the conversion of nutrients into building blocks (intermediary metabolism), signaling, and apoptosis. One of the most exciting areas of research involves the regulation of nuclear gene expression by mitochondria through "retrograde communication." While this pathway is well-defined in yeast, it lacks conserved components in higher eukaryotes and how the expression of nuclear-encoded mitochondrial genes is adjusted to metabolic requirements and respiratory activity in mammalian cells is under intense investigation.[6,7] Apart from activation or silencing of specific genes, mitochondria also regulate global transcription through metabolites that influence the epigenetic landscape (Section 5).

Mitochondrial diseases have classically been defined as based on mutations of the mitochondrial genome. However, the majority of the more than 1000 mitochondrial proteins is encoded in the nucleus and not inherited strictly maternally.[8] The list of diseases that are based on nuclear-encoded mitochondrial disorders is growing, as is our understanding of the involvement of this organelle in basic and complex cellular functions. In addition to metabolic dysfunction caused directly by mitochondrial impairment, several diseases are also indirectly influenced by mitochondrial activity, for instance neurodegenerative diseases and cancer. Given their dominant role in metabolism, it is not surprising that mitochondria also play a central role in the aging process.[9–11]

2.1 ROS Theory of Aging

One of the most widely published theories of aging involves reactive oxygen species (ROS).[12] These molecules are by-products of the respiratory chain in the inner mitochondrial membrane, where redox reactions create a proton gradient that drives ATP generation. Superoxide radicals are produced by complex (I) and complex (III) of the respiratory chain. Three scavenging mechanisms, superoxide dismutase [(SOD); mitochondrion and cytosol], catalase (peroxisome), and glutathione peroxidase (mitochondrion and cytosol) act in a stepwise fashion to remove these radicals. First, SOD detoxifies the superoxide and creates hydrogen peroxide. This intermediary can form the more dangerous hydroxyl radical if not removed. Catalase and glutathione peroxidase prevent hydroxyl radical formation by converting hydrogen peroxide into water. Increased SOD activity without changes to catalase or glutathione peroxidase can be harmful due to accumulation of damaging hydroxyl radicals. In contrast, elevated levels of glutathione promote glutathione peroxidase's reducing activity and can have beneficial health effects. Patients who were fed bioavailable precursors of glutathione had significantly improved laboratory and clinical parameters in the HIV model of accelerated aging.[13–15] Studies are underway to test whether increasing glutathione through food supplementation with the precursor amino acids cysteine and glycine will improve health or life span in this clinical model.

Increased ROS production through desynchronized respiration and/or lower ROS scavenging capacity with age offers an attractive model to explain damage to proteins through carbonylation or to DNA through guanine modifications.[16,17] With their proximity to the respiratory chain, mitochondrial proteins and mitochondrial DNA that lacks protective nucleosomes are especially prone to damage. Indeed, mice with error-prone mitochondrial DNA replication age prematurely.[18,19] However, the theory of ROS-induced aging has been reevaluated over the past decade. Genetic alterations in ROS scavengers or treatment with antioxidants failed to show consistent effects on life span regulation.[20] Different species' life expectancies do not correlate with ROS levels. Moreover, there is evidence that modest levels of ROS have an antiaging effect by triggering a stress-relieving response (hormesis). Combined, it appears that the dosage is critical when evaluating ROS in the context of aging, but reduction of ROS alone does not extend life span under physiological conditions.

2.2 Mitochondrial Aging

Mitochondria replicate independently of the cell cycle. In addition, they undergo fusion–fission cycles, which serve to adapt ATP production levels, allow for repair of mitochondrial DNA damage, and adjust the level of heteroplasmy (i.e., the degree to which multiple different mitochondrial DNA species coexist in the same cell).[21] Mitochondrial removal by autophagy also requires fission of the organelle (Section 3). Aged cells have impaired fusion and fission, thus reducing overall metabolic flexibility. As a consequence, mitochondria fail to adapt to nutrient availability or signaling cues and cannot switch efficiently between utilization of the substrates fatty acids, glucose, and amino acids.[22]

There is solid evidence that mitochondrial activity declines with age in tissues, as well as in individual cells.[23] Well documented types of damage include replication errors of mitochondrial DNA and toxic adducts on mitochondrial proteins. The elevated pH and high levels of acetyl-CoA and succinyl-CoA in the mitochondrial matrix promote nonenzymatic acylation of ε-amino groups. Such modifications of the amino acid lysine can prove harmful to the function of proteins. The removal of these adducts is catalyzed by sirtuins, a class of protein deac(et)ylases that are particularly enriched in mitochondria (SIRT3, SIRT4, and SIRT5). Their activity improves mitochondrial function and contributes to healthy aging, though not necessarily extending life expectancy.[24] Sirtuins require NAD^+ as coenzyme and this metabolite is a key antiaging effector (Section 4). In addition to the mitochondrial sirtuins, SIRT1 promotes mitochondrial activity by deacetylating the transcriptional coactivator PGC1α, allowing it to promote the expression of NRF-1 target genes. NRF-1 is a master transcription factor of numerous nuclear-encoded mitochondrial genes. PGC1α, together with increasing levels of ROS, also stimulates NRF-2, which activates a support pathway and increases the expression of antioxidant stress response genes.[25] Combined, sirtuins increase mitochondrial activity and preserve their function by adapting the stress response (Fig. 1). SIRT1 and SIRT3 in particular have been linked to maintain metabolic health, though initial findings on their ability to extend life span have been disproven.[26,27]

3. PROTEOSTASIS NETWORK

In many cells, protein synthesis and removal are the most resource-consuming processes, tightly connecting them to energy metabolism. Furthermore, proteins are the main executors of cell function and dysregulated production or protein dysfunction have been implicated in aging. In particular, the highly specialized neurons of the mammalian brain seem to be sensitive to protein damage.[28] The expression of aggregation-prone polypeptides and the inability of these cells to divide and dilute protein accumulation are the root causes of several neurodegenerative diseases. Protein stress by itself may not damage neurons, but makes these frail cells more susceptible

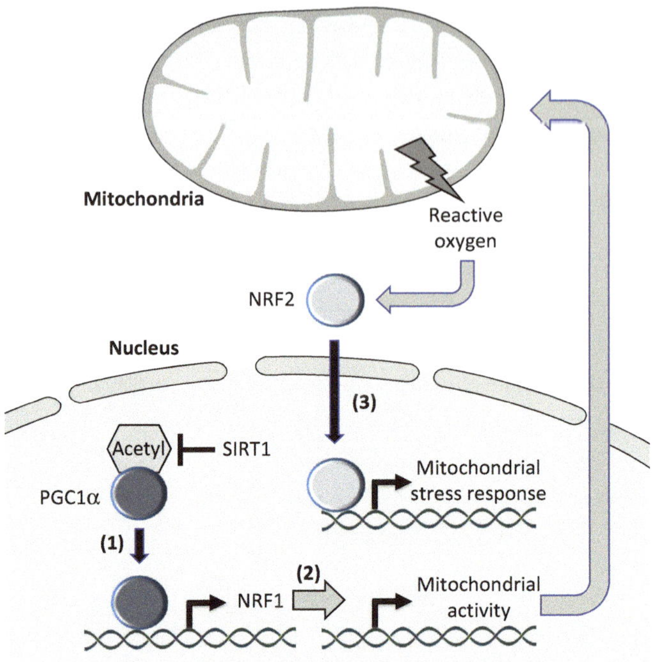

Fig. 1 Master regulators of mitochondrial activity. SIRT1 activates expression of nuclear-encoded mitochondrial genes through deacetylation of PGC1α. The sirtuins SIRT3, SIRT4, and SIRT5 reduce proteotoxic stress by removing protein acetylation in mitochondria (not shown).

to a second toxic injury, such as inflammation or metabolic insults. Apart from neurons, most cells show consistent changes to protein homeostasis (proteostasis) with age, such as impaired proteasome and chaperone function.[29–34]

3.1 Translation Control

Proteostasis involves protein synthesis, structural and topological changes, and polypeptide removal. Lowering the biogenesis of polypeptides indeed alleviates stress and inhibition of global translation has positive effects on life span extension.[35] Experimental reduction of translation includes genetic manipulation of ribosomes or translation factors, as well as pharmacological interventions, most notably through inhibition of the mTOR signaling pathway (Section 4). Cell stress can also reduce overall protein synthesis. Attenuated translation is accomplished by phosphorylation of the translation initiation factor eIF2α by kinases, such as GCN2 and PKR. However, select proteins may be translated at higher rates under these conditions as their synthesis is either independent of eIF2α or even inhibited by native eIF2α.[36] Among them is the transcription factor ATF4 that drives the integrated stress response (ISR) that favors expression of life span-extending factors, such as transcription factors and executors dealing with oxidative damage. This longevity-promoting effect of the ISR is also triggered by caloric restriction, the most successful intervention thus far to improve longevity and health span.

3.2 UPR

The main purpose of the unfolded protein response (UPR) is to reduce protein stress in the endoplasmic reticulum (ER) by altering protein production in a manner that is similar and partially overlapping with the ISR (Fig. 2).[37] Triggered by misfolding of proteins within the ER, the UPR attenuates global translation through phosphorylation of initiation factor eIF2α by the ER-resident kinase PERK. A second arm of the UPR involves nuclear translocation of the transcription factor ATF6, which increases expression of ER-resident chaperones. Third, activation of the RNase IRE1 leads to alternative splicing of the transcription factor XBP1, which alters the proteostatic milieu in the ER.[38] Slight activation of the UPR, in particular higher expression of XBP1, has beneficial effects on survival and life span.[37] Interestingly, XBP1 also acts in a cell nonautonomous manner, as a stress regulator in the nervous system of nematodes. Upregulation of XBP1 in neurons leads to UPR activation in peripheral organs.[39]

3.3 Chaperone Network

Protein synthesis requires 30%–50% of cellular energy and is error-prone even under optimal conditions. By some estimates, a third of polypeptides fail to fold properly during translation in mammalian cells.[40–43] A network of chaperones supports protein folding and activity from translation to degradation. Initially, ribosome-associated chaperones and foldases ensure proper structural assembly of a newly synthesized polypeptide. The nascent polypeptide-associated complex (NAC) accounts for most of the chaperone-assisted protein folding of newly translated proteins.[44] A second protective layer is provided by chaperones, such as HSP70 and HSP40, which fold proteins in an ATP-dependent manner. Finally, if nascent proteins require additional assistance, they are transferred into chaperonins, closed nanocages

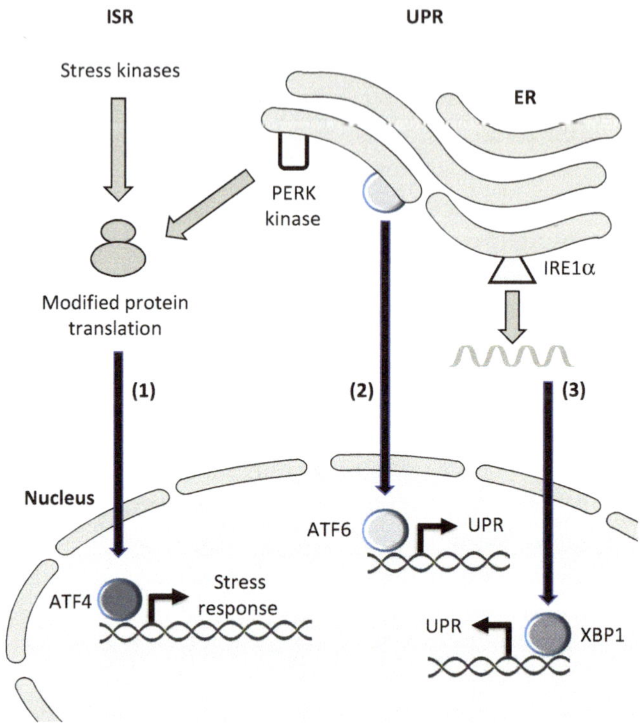

Fig. 2 Integrated Stress Response (ISR) and unfolded protein response (UPR). Upstream stress kinases, such as GCN2 and PKR, as well as the endoplasmic reticulum (ER)-bound kinase PERK, phosphorylate eIF2α (1). This reduces overall protein translation, while favoring the expression of the transcription factor ATF4 and several chaperones. Additional elements of the UPR include the transcription factor ATF6, which translocates from the ER to the nucleus upon stress (2), and IRE1α, which allows for alternative splicing of the mRNA encoding the transcription factor XBP1, leading to its activation (3).

that allow protein folding in isolation and without interference from the otherwise crowded protein environment of the cytosol.[45] Consistent failure to fold proteins eventually leads to their removal through the ubiquitin-proteasome system (UPS) or autophagy (see further).

Sentinel chaperones, such as HSP90 scan the mature proteome for misfolded polypeptides. These chaperones can attempt to refold client proteins or direct them for removal. Approximately 332 chaperones and associated proteins are encoded in the human genome.[46] This "chaperome" consists of constitutive elements but is also enriched for stress response factors that are upregulated by heat shock or by the transcription factors ATF4 (for the ISR) and ATF6 (for the UPR). The dynamic response to protein stress is reduced with age and transgenic overexpression of chaperones improves life span.[47–49] However, it should be noted that abnormally high levels of chaperones come at a cost: they dampen signaling by the heat shock transcription factor HSF1 through feedback mechanisms and eliminate the ability to mount a proper stress response.[50] Further, overexpressed chaperones can soak up client polypeptides and thus reduce the bioavailability of critical proteins.

3.4 UPS

The UPS is the primary pathway to eliminate proteins in a specific fashion. A decrease of UPS activity with age has been linked to deficiencies in proteostasis. Further, one of the functions that distinguishes the long-lived naked mole rat from shorter lived rodents is the robustness of its UPS.[51,52] Similarly, healthy centenarians have been shown to maintain high proteasomal activity.[53]

Proteins that are destined for degradation by the proteasome are first tagged with several copies of the small polypeptide ubiquitin in a chain-like fashion. The covalent attachment of ubiquitin to target proteins occurs through a series of enzymatic events, involving ubiquitin-activating enzymes, conjugating enzymes, and finally ubiquitin ligases. Approximately 600 ubiquitin ligases exist in human cells. These ligases can associate with hundreds of different substrate recognition domains, allowing for a substantial combinatorial diversity that ensures specificity of the tagging process.[29] Similar to the chaperone network, several proteins of the UPS are upregulated upon stress. For instance, the UPR activates ubiquitination enzymes that allow for the specific extraction, ubiquitination, and proteasomal degradation of ER-resident proteins in a process called endoplasmic reticulum-associated protein degradation (ERAD). The UPS also regulates chaperone activity with ubiquitin ligases, such as CHIP (also known as STUB1), which tags chaperone client proteins for degradation during stress and ubiquitinates the chaperone HSP70 for removal during recovery from stress.[54] The ability of the UPS to target specific substrates is also utilized for the removal of organelles or bulk protein aggregates through autophagy (see further). Here, ubiquitination serves as a ligand for autophagosome receptors.

The UPS is a critical regulator of gene expression through the removal of short-lived transcription factors. Degradation therefore has wide-ranging effects on cellular activity and metabolism. One example of metabolic feedback mechanisms between mitochondria, the UPS, and gene activity involves ROS. Chemical radicals can dampen the activity of the UPS through carbonylation of the proteasome, which in turn can stabilize otherwise short-lived transcriptional repressors. This reduces the expression of nuclear-encoded mitochondrial genes and the generation of ROS. Replacement of carbonylated proteasome subunits through new rounds of transcription and translation, and the activation of transcription factor NRF-2, which leads to the production of ROS scavengers and executors of autophagy and the UPS, eventually normalize oxidative metabolism following ROS exposure.[29,55,56]

3.5 Mitochondrial Proteostasis

Mitochondria have a select number of proteases and chaperones that are expressed in the cytosol and imported into the intermembrane or matrix chambers, or anchored to the membranes. These mitochondrial proteases ensure the import and activity of proteins into mitochondria, and maintain the homeostasis of the respiratory chain complex, which is partly encoded by the mitochondrial genome. Protein stress inside mitochondria triggers a mitochondrial version of the UPR (UPR^{mt}), which increases expression of chaperones, proteases, and ROS detoxifying factors.[57] The UPR^{mt} is responsible for refolding or destroying misfolded proteins and thereby improving the efficiency and the health of mitochondria. Notably, a recent publication pointed out this stress response can also have deleterious effects on mitochondrial homeostasis. When healthy and damaged mitochondria coexist (heteroplasmy), the UPR^{mt} can provide sufficient buffering capacity to maintain dysfunctional organelles, which, in the absence of UPR^{mt}, would have been competitively replaced by healthy mitochondria.[58] Regarding aging, the role of the UPR^{mt} remains contested. Upregulation of some UPR^{mt} elements promotes resilience to ROS and increases life span (mitohormesis), while other elements accomplish the opposite. More work will be necessary before a verdict can be reached on impact the UPR^{mt} has on aging.[59,60]

Crosstalk between the mitochondrial and cytosolic proteostasis systems occurs at multiple layers and may also play a role in the aging process (Fig. 3). A dedicated cytosolic stress response deals with mitochondrial proteins that have failed to enter the organelle. Further, mitochondria-associated degradation extracts proteins from the organelle for cytosolic degradation by the

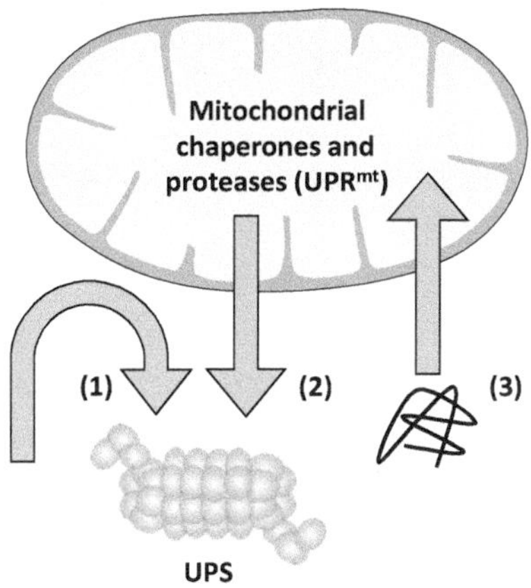

Fig. 3 Cooperation between mitochondrial proteostasis and the ubiquitin-proteasome system (UPS). Mitochondrial proteins that fail to import into the organelle are removed by the UPS (1). Misfolded proteins can be extracted from mitochondria for proteasomal degradation; similar to the way, misfolded proteins are extracted from the ER (2). Protein aggregates can be imported by mitochondria and digested by resident proteases (3).

UPS, in a mechanism resembling the UPR.[61,62] In addition, mitochondrial dysfunction can elicit a cytosolic stress response that enhances nonmitochondrial proteostasis and increases cellular fitness.[63]

Mitochondria have also been suggested to directly assist cytosolic proteostasis by importing protein aggregates for digestion by matrix proteases. This observation might explain why mitochondrial defects impact some neurodegenerative diseases, not only through weakening the metabolism of neurons but also by failing to clear protein aggregates. Indeed, mitochondrial dysfunction accelerates the course of some protein aggregate diseases.[64] How these aggregated proteins mechanistically enter mitochondria and how this pathway affects aging and disease awaits further clarification.[65]

3.6 Autophagy

Autophagy is the proteostatic mechanism with the strongest ties to longevity. In simpler eukaryotes, life span extension by rapamycin requires induction of autophagy.[66,67] In addition, direct overexpression of regulators and executors of autophagy have been shown to have beneficial effects on health and life span.[68] Autophagy can be divided into three subtypes. First, chaperone-mediated autophagy targets individual proteins through interaction with heat shock proteins for lysosomal import and degradation. Second, microautophagy involves direct engulfment of proteins by the lysosome. The third, macroautophagy, is the most important mechanism for metabolic control and life span regulation.[69] Following initiation of the phagosome at mitochondrial-ER contact surfaces, a growing vesicle engulfs macromolecular complexes or organelles and eventually fuses to the lysosome for degradation. Macroautophagy is important for the recycling of lipids, sugars, amino acids, and metals, and its contribution to life span extension is based on the removal of damaged proteins, superfluous lipids, impaired organelles, and the regulation of inflammation and cell death.

An important aspect of macroautophagy is the removal of damaged or excess mitochondria. Mitochondrial dysfunction, for instance, reduces the ability of these organelles to import proteins. The kinase PINK1 acts as an indicator of import stress and accumulates on the outer membrane. PINK1 phosphorylates ubiquitin and activates the ubiquitin ligase PARKIN. PARKIN conjugates a plethora of accessible mitochondrial proteins with phosphoubiquitin, which attracts phagosome receptors for internalization of mitochondria (mitophagy).[70,71] Mutations in this pathway lead to early onset Parkinson's disease. The control of mitochondrial numbers and the overall state of aerobic metabolism of cells is also regulated through mitophagy. One study exemplified how aged hematopoietic stem cells with increased autophagy display healthier, more juvenile features. The authors argued that removal of mitochondria and shifting of the overall metabolism toward glycolysis contributed to increased cellular health.[72] Apart from the engulfment of entire mitochondria, smaller portions can also be pinched off in vesicles containing oxidized proteins for lysosomal digestion in a process that is independent from autophagy or the fusion/fission cycle. These carriers are referred to as mitochondria-derived vesicles.[73]

The scale of autophagy has been studied best in the liver, where up to 70% of proteome turnover occurs through this mechanism. Inhibition of autophagy leads to the accumulation of poly-ubiquitinated inclusions, indicating that the proteasome system alone is insufficient for bulk protein removal. Instead, the UPS functions mainly in selective polypeptide degradation or in ubiquitination as a recognition tag to trigger autophagy. Autophagy has also been implicated in restructuring of the nuclear architecture. However, it remains to be seen whether nuclear autophagy is involved in the aging process.[74,75]

Lysosomes are the endpoint of phagosome breakdown and have remarkable capacity to extract and recycle individual nutrients and building blocks.[76] Recently, the role of lysosomes in the aging process has been extended to a lysosome-to-nucleus signaling function and additional lysosomal pathways with relevance for aging are likely awaiting discovery.[77]

4. NUTRIENT-SENSING PATHWAYS

Pathways regulating energy metabolism and adaptation to nutrients are commonly associated with life span control. The main players are the insulin-IGF1 pathway, mTOR-containing complexes, AMPK, and sirtuins (Fig. 4). The mechanisms of these pathways have been covered in detail in a series of excellent reviews.[3,4,78–83] Here, we will briefly outline their function and their interplay with other cellular systems pertaining to life span regulation.

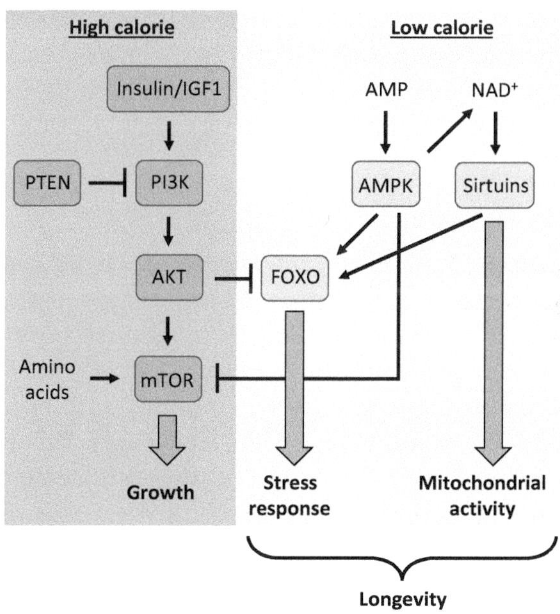

Fig. 4 Nutrient-sensing pathways. Overview of the main pathways that sense sugar levels (insulin/IGF1) and amino acids (mTOR) or loss of bioavailable energy (AMPK) and NAD^+ depletion (sirtuins). High calorie conditions favor growth, while low calorie conditions improve the stress response, optimize energy metabolism, and have the potential to increase the life span.

4.1 Growth Pathways

The insulin-IGF1 pathway regulates growth in times of ample nutrient supply. Through intermediaries, such as PI3K and Akt, this pathway suppresses FOXO transcription factors. FOXO transcription factors have multiple downstream effects: they activate autophagy, the UPS, and the ROS stress response, are essential for stem cell function, have tumor-suppressive activities, and promote life span extension.[84] Phosphorylation by Akt prohibits nuclear translocation of FOXOs and instead activates the mTOR complex.[85] The mTOR kinase complex is a master regulator of cell growth and mTOR activation increases protein synthesis, energy expenditure, and induces a metabolic switch from frugality to anabolic processes that accelerate aging. The insulin-IGF1 pathway responds to high glucose levels, while mTOR responds directly to high amino acid concentrations and indirectly to glucose levels through Akt. To date, the mTOR inhibitor rapamycin has shown the most promising pharmacologic effect on life span extension in mammals. As mentioned earlier, the positive impact of rapamycin is autophagy-dependent. The other highly effective intervention, caloric restriction, slows cell growth by reducing both insulin-IGF1 and mTOR signaling.[80] However, the life span extension mediated by these interventions is limited and sustained inhibition of vital systems can become harmful, increase insulin resistance, lower cellular fitness, and accelerate aging over time.[3]

4.2 Restrictive Pathways

Whereas insulin-IGF1 and mTOR represent anabolic "high energy" pathways, the catabolic "low energy" factors AMPK and sirtuins[86,87] counteract mTOR and enhance mitochondrial activity to produce more ATP through fatty acid oxidation, and induce autophagy to recycle nutrients.[88–90] Unfortunately, the westernized lifestyle, consisting of a hypercaloric diet and limited exercise, downregulates AMPK and sirtuins. Caloric restriction, on the other hand, activates both pathways, stimulating efficient ATP production through glucose uptake and glycolysis, fatty acid oxidation, and inducing recycling pathways to replenish energy.[91,92] AMPK is a structural sensor of the AMP/ATP ratio and low energy (high AMP) allows AMPK to be phosphorylated by its kinase LKB1. Pharmacologic activation of AMPK can be achieved by the antidiabetic drug metformin, which will be used in the first life span-extension study in humans [targeting aging with metformin (TAME)].[4,93] As a kinase, AMPK phosphorylates a number of target proteins to increase catabolic metabolism and inhibit mTOR. AMPK is a key mediator of longevity and induction of autophagy is an important downstream effect.[94] The recycling of cellular organelles increases lysosomal amino acid levels, which are detected by the mTOR complex. Amino acid-induced activation of mTOR then terminates the catabolic reaction in a homeostatic feedback loop.

Sirtuins are NAD^+-dependent deac(et)ylases that exert several beneficial effects, such as detoxifying mitochondria predominantly via SIRT3 (Section 2), increasing genomic stability through chromatin-associated SIRT6, and transcriptional activation of mitochondria by SIRT1.[24,95,96] The additional function of sirtuins as ADP-ribosyltransferases is outside the scope of this review. With their dependence on NAD^+, sirtuins effectively act as sensors for this coenzyme.[95] NAD^+ synthesis is under circadian control and can be quickly increased by fasting and AMPK activation. Importantly, NAD^+ levels are depleted with age, partially through competition between sirtuins and other enzymes, such as the NAD^+-dependent DNA repair protein PARP.[4] Increased NAD^+ supply, on the other hand, has been shown to improve health and life span. Some of the immediate metabolic changes following NAD^+ repletion involve enhanced mitochondrial activity through deacetylation and activation of PGC1α by SIRT1, as well as improved mitochondrial proteostasis by SIRT3. Recently, SIRT7 has also been shown to improve mitochondrial health and aging in hematopoietic stem cells.[97]

5. COORDINATION OF NUCLEAR ACTIVITIES AND METABOLISM

5.1 The Nuclear-Metabolic Axis

Disruption of the nuclear architecture can have profound effects on aging, as shown by progeria-causing laminopathies. These diseases cause juvenile-onset aging, as in Hutchinson–Gilford progeria syndrome. Similar to DNA repair defects, affected patients age at an accelerated pace and die prematurely.[98,99] Laminopathies interfere with autophagy and impede mitochondrial activity through only partially understood mechanisms.[100] The mTOR inhibitor rapamycin has been shown to improve the cellular function of progeroid fibroblasts and treatment protocols are under development for these currently incurable diseases.[101]

Several pathways linked to cancer, specifically cell cycle control genes $p16^{INK4a}$, $p19^{ARF}$, and the tumor suppressor p53, also play important roles in the aging process.[3,102–104] The expression levels of p16^{INK4a} and p19^{ARF} correlate with age and using them as chronological markers has been proposed. Polymorphisms in these genes are associated with age-related diseases. Mild increases of p16^{INK4a}, p19^{ARF}, and p53 have beneficial effects on aged cells, independent of their tumor-suppressive effects, but more significant upregulation accelerates aging. The mechanistic basis of this phenomenon remains unclear.

Attrition of telomeres is a consequence of cell replication and aging. Shortened telomeres induce senescence and trigger a p53-dependent response that shuts down mitochondrial activity.[105] However, p53 can also increase mitochondrial output through association with the mitochondrial transcription factor TFAM in the matrix.[106] Therefore, key regulators of the cell cycle appear to have varied roles in metabolism and aging, depending on the context. Assembling a unifying model that explains all these activities poses a major challenge in this field.

5.2 The Metabolic-Nuclear Axis

Studies on cancer metabolism and aging have converged on the link between the intermediary metabolism and epigenetics. DNA- and histone-modifying enzymes are key modifiers of the epigenetic landscape and require intermediary metabolites. There is growing interest in dissecting how metabolism affects gene expression and vice versa. One hallmark of aged cells is the loss of gene silencing.[107] Increased transcriptional noise exists due to a lack of heterochromatin, reactivation of transposons, dysregulated splicing, and other impairments.[108–111] In addition, epigenetic modifications can promote the inheritance of transgenerational longevity and metabolic traits in model organisms, as well as in humans.[112–114] Among the epigenetic changes that occur during aging are loss of H3K9me3 modifications (a marker of gene repression), global loss of DNA methylation, and enhanced H4K16 acetylation (a marker of active genes).[115–117] How these alterations are affected by the overall metabolic state and how they in turn regulate cell function is not entirely understood. Still, distinct metabolic pathways generate intermediates that are required for the enzymatic reactions that add or remove methyl or acetyl residues.[118–121]

The metabolites S-adenosyl methionine and acetyl-CoA are required for DNA and histone methylation, and for protein acetylation, respectively. Both are produced under nutrient-rich conditions. Their antagonists are α-ketoglutarate, which is required for protein and DNA demethylation and

NAD^+, a coenzyme of sirtuin-dependent protein deacetylation. Both intermediates can increase under nutrient-low conditions: α-ketoglutarate levels are replenished by metabolic reactions that feed into the TCA cycle (anaplerosis) and NAD^+ availability is regulated through AMPK-dependent synthesis.[122,123] While caloric restriction and elevated NAD^+ may attenuate aging by reducing protein acetylation, the effects on histone acetylation and gene activity are less obvious and await further clarification.[124,125]

NAD^+ also serves as a coenzyme of the DNA repair enzyme PARP and NAD^+ loss has been linked to genomic damage with age.[126] The TCA cycle intermediate α-ketoglutarate displays metabolic fluctuations. Hypoxia, for instance, favors the generation of its antagonist 2-hydroxy-glutarate (L2HG), which inhibits demethylase activity.[127,128] Formation of 2HG is also enhanced in certain cancers with mutations of the TCA enzyme IDH. These IDH mutations convert α-ketoglutarate into R2HG, which like L2HG blocks demethylation.[129]

Epigenetic changes during aging affect the genome globally, but also locally. For instance, some genes defy global trends and are under direct regulation by metabolic pathways, such as FOXO target genes or stress response genes that are induced by ROS through the transcription factor NRF-2. Further, the cell cycle, as well as senescence can be regulated by metabolism and mitochondrial activity through p53.[130,131] For instance, under starvation AMPK phosphorylates p53 to induce p21, causing a cell cycle arrest. Mitochondrial stress in form of increased ROS levels also induces cell cycle arrest using mediators, such as p53 and p27.[130] Mitochondria can stimulate the expression of nuclear UPR^{mt} genes through alterations in the epigenetic code.[132] It has been shown in worms that mitochondrial dysfunction increases histone demethylases to promote these UPR^{mt} factors. Interestingly, to extend life span this effect needs to occur in the nervous system. The complex and sometimes contradictory effects of global and local epigenetic changes, as well as their connections with metabolismand aging defy summarization in a simple model.[133]

6. SUMMARY

Genome-wide genetic studies in yeast and worms have led to the discovery of the major pathways that govern longevity. Hormesis is a common principle across systems that applies to several of the described longevity pathways: limited metabolic impairment extends life span due to compensatory stress responses, whereas too much damage becomes harmful. Research findings based on blunt knock-out experiments may therefore require reevaluation in systems that allow for gene dosage adjustments. Still, the most exciting aspect of metabolism remains in the interconnectivity of pathways. How these different systems communicate with each other to synchronize their activity is the target of ongoing studies. Indeed, we are likely to make the most exciting discoveries about metabolic regulation of aging and find new avenues for therapeutic intervention at the intersection of these pathways.

ACKNOWLEDGMENTS

This review was supported by the Cancer Prevention and Research Institute of Texas (RR140038), the Ted Nash Long Life Foundation, and NIH R01DK115454. We thank Catherine Gillespie for helpful comments and suggestions.

REFERENCES

1. Cai W, et al. Oral glycotoxins are a modifiable cause of dementia and the metabolic syndrome in mice and humans. *Proc Natl Acad Sci USA*. 2014;111:4940–4945.
2. Hayflick L. Recent advances in the cell biology of aging. *Mech Ageing Dev.* 1980;14:59–79.
3. Lopez-Otin C, Blasco MA, Partridge L, Serrano M, Kroemer G. The hallmarks of aging. *Cell.* 2013;153:1194–1217.
4. Lopez-Otin C, Galluzzi L, Freije JM, Madeo F, Kroemer G. Metabolic control of longevity. *Cell.* 2016;166:802–821.
5. Bjorkholm P, Harish A, Hagstrom E, Ernst AM, Andersson SG. Mitochondrial genomes are retained by selective constraints on protein targeting. *Proc Natl Acad Sci USA.* 2015;112:10154–10161.
6. Liu Z, Butow RA. Mitochondrial retrograde signaling. *Annu Rev Genet.* 2006;40:159–185.
7. Quiros PM, Mottis A, Auwerx J. Mitonuclear communication in homeostasis and stress. *Nat Rev Mol Cell Biol.* 2016;17:213–226.
8. Calvo SE, Mootha VK. The mitochondrial proteome and human disease. *Annu Rev Genomics Hum Genet.* 2010;11:25–44.
9. Bratic A, Larsson NG. The role of mitochondria in aging. *J Clin Invest.* 2013;123:951–957.
10. Rose G, Santoro A, Salvioli S. Mitochondria and mitochondria-induced signalling molecules as longevity determinants. *Mech Ageing Dev.* 2017;165:115–128.
11. Sack MN, Finkel T. Mitochondrial metabolism, sirtuins, and aging. *Cold Spring Harb Perspect Biol.* 2012;4:1–10.
12. Hohn A, Konig J, Grune T. Protein oxidation in aging and the removal of oxidized proteins. *J Proteomics.* 2013;92:132–159.
13. Sekhar RV, Liu CW, Rice S. Increasing glutathione concentrations with cysteine and glycine supplementation lowers inflammation in HIV patients. *Aids.* 2015;29:1899–1900.
14. Nguyen D, Hsu JW, Jahoor F, Sekhar RV. Effect of increasing glutathione with cysteine and glycine supplementation on mitochondrial fuel oxidation, insulin sensitivity, and body composition in older HIV-infected patients. *J Clin Endocrinol Metab.* 2014;99:169–177.

15. Pathai S, Bajillan H, Landay AL, High KP. Is HIV a model of accelerated or accentuated aging? *J Gerontol A Biol Sci Med Sci.* 2014;69:833–842.
16. Jena NR. DNA damage by reactive species: mechanisms, mutation and repair. *J Biosci.* 2012;37:503–517.
17. Suzuki YJ, Carini M, Butterfield DA. Protein carbonylation. *Antioxid Redox Signal.* 2010;12:323–325.
18. Trifunovic A, et al. Premature ageing in mice expressing defective mitochondrial DNA polymerase. *Nature.* 2004;429:417–423.
19. Kujoth GC, et al. Mitochondrial DNA mutations, oxidative stress, and apoptosis in mammalian aging. *Science.* 2005;309:481–484.
20. Perez VI, et al. The overexpression of major antioxidant enzymes does not extend the lifespan of mice. *Aging Cell.* 2009;8:73–75.
21. Chauhan A, Vera J, Wolkenhauer O. The systems biology of mitochondrial fission and fusion and implications for disease and aging. *Biogerontology.* 2014;15:1–12.
22. Muoio DM. Metabolic inflexibility: when mitochondrial indecision leads to metabolic gridlock. *Cell.* 2014;159:1253–1262.
23. de Moura MB, dos Santos LS, Van Houten B. Mitochondrial dysfunction in neurodegenerative diseases and cancer. *Environ Mol Mutagen.* 2010;51:391–405.
24. Someya S, et al. Sirt3 mediates reduction of oxidative damage and prevention of age-related hearing loss under caloric restriction. *Cell.* 2010;143:802–812.
25. Scarpulla RC. Nuclear control of respiratory chain expression by nuclear respiratory factors and PGC-1-related coactivator. *Ann N Y Acad Sci.* 2008;1147:321–334.
26. Burnett C, et al. Absence of effects of Sir2 overexpression on lifespan in *C. elegans* and *Drosophila*. *Nature.* 2011;477:482–485.
27. Viswanathan M, Guarente L. Regulation of *Caenorhabditis elegans* lifespan by sir-2.1 transgenes. *Nature.* 2011;477:E1–E2.
28. Saxena S, Caroni P. Selective neuronal vulnerability in neurodegenerative diseases: from stressor thresholds to degeneration. *Neuron.* 2011;71:35–48.
29. Dikic I. Proteasomal and autophagy degradation systems. *Annu Rev Biochem.* 2017;86:193–224.
30. Balchin D, Hayer-Hartl M, Hartl FU. In vivo aspects of protein folding and quality control. *Science.* 2016;353:aac4354.
31. Garcia-Prat L, Sousa-Victor P, Munoz-Canoves P. Proteostatic and metabolic control of stemness. *Cell Stem Cell.* 2017;20:593–608.
32. Sadowska-Bartosz I, Bartosz G. Prevention of protein glycation by natural compounds. *Molecules.* 2015;20:3309–3334.
33. Sala AJ, Bott LC, Morimoto RI. Shaping proteostasis at the cellular, tissue, and organismal level. *J Cell Biol.* 2017;216:1231–1241.
34. Vilchez D, Simic MS, Dillin A. Proteostasis and aging of stem cells. *Trends Cell Biol.* 2014;24:161–170.
35. Steffen KK, Dillin A. A ribosomal perspective on proteostasis and aging. *Cell Metab.* 2016;23:1004–1012.
36. Holcik M. Could the eIF2alpha-independent translation be the Achilles heel of cancer? *Front Oncol.* 2015;5:264.
37. Martinez G, Duran-Aniotz C, Cabral-Miranda F, Vivar JP, Hetz C. Endoplasmic reticulum proteostasis impairment in aging. *Aging Cell.* 2017;16:615–623.
38. Shoulders MD, et al. Stress-independent activation of XBP1s and/or ATF6 reveals three functionally diverse ER proteostasis environments. *Cell Rep.* 2013;3:1279–1292.
39. Taylor RC, Dillin A. XBP-1 is a cell-nonautonomous regulator of stress resistance and longevity. *Cell.* 2013;153:1435–1447.
40. Glembotski CC. Endoplasmic reticulum stress in the heart. *Circ Res.* 2007;101: 975–984.

41. Buttgereit F, Brand MD. A hierarchy of ATP-consuming processes in mammalian cells. *BiochemJ.* 1995;312(Pt 1):163–167.
42. Russell JB, Cook GM. Energetics of bacterial growth: balance of anabolic and catabolic reactions. *Microbiol Rev.* 1995;59:48–62.
43. Schubert U, et al. Rapid degradation of a large fraction of newly synthesized proteins by proteasomes. *Nature.* 2000;404:770–774.
44. Hartl FU. Protein misfolding diseases. *Annu Rev Biochem.* 2017;86:21–26.
45. McGuffee SR, Elcock AH. Diffusion, crowding & protein stability in a dynamic molecular model of the bacterial cytoplasm. *PLoS Comput Biol.* 2010;6:e1000694.
46. Brehme M, et al. A chaperome subnetwork safeguards proteostasis in aging and neurodegenerative disease. *Cell Rep.* 2014;9:1135–1150.
47. Calderwood SK, Murshid A, Prince T. The shock of aging: molecular chaperones and the heat shock response in longevity and aging–a mini-review. *Gerontology.* 2009;55:550–558.
48. Morrow G, Samson M, Michaud S, Tanguay RM. Overexpression of the small mitochondrial Hsp22 extends *Drosophila* life span and increases resistance to oxidative stress. *FasebJ.* 2004;18:598–599.
49. Walker GA, Lithgow GJ. Lifespan extension in *C. elegans* by a molecular chaperone dependent upon insulin-like signals. *Aging Cell.* 2003;2:131–139.
50. Voellmy R. On mechanisms that control heat shock transcription factor activity in metazoan cells. *Cell Stress Chaperones.* 2004;9:122–133.
51. Perez VI, et al. Protein stability and resistance to oxidative stress are determinants of longevity in the longest-living rodent, the naked mole-rat. *Proc Natl Acad Sci USA.* 2009;106:3059–3064.
52. Chondrogianni N, et al. Protein damage, repair and proteolysis. *Mol Aspects Med.* 2014;35:1–71.
53. Chondrogianni N, Petropoulos I, Franceschi C, Friguet B, Gonos ES. Fibroblast cultures from healthy centenarians have an active proteasome. *Exp Gerontol.* 2000;35:721–728.
54. Qian SB, McDonough H, Boellmann F, Cyr DM, Patterson C. CHIP-mediated stress recovery by sequential ubiquitination of substrates and Hsp70. *Nature.* 2006;440:551–555.
55. Segref A, et al. Pathogenesis of human mitochondrial diseases is modulated by reduced activity of the ubiquitin/proteasome system. *Cell Metab.* 2014;19:642–652.
56. Catic A, et al. Genome-wide map of nuclear protein degradation shows NCoR1 turnover as a key to mitochondrial gene regulation. *Cell.* 2013;155:1380–1395.
57. Voos W, Rottgers K. Molecular chaperones as essential mediators of mitochondrial biogenesis. *Biochim Biophys Acta.* 2002;1592:51–62.
58. Lin YF, et al. Maintenance and propagation of a deleterious mitochondrial genome by the mitochondrial unfolded protein response. *Nature.* 2016;533:416–419.
59. Bennett CF, Kaeberlein M. The mitochondrial unfolded protein response and increased longevity: cause, consequence, or correlation? *Exp Gerontol.* 2014;56:142–146.
60. Tian Y, Merkwirth C, Dillin A. Mitochondrial UPR: a double-edged sword. *Trends Cell Biol.* 2016;26:563–565.
61. Wrobel L, et al. Mistargeted mitochondrial proteins activate a proteostatic response in the cytosol. *Nature.* 2015;524:485–488.
62. Taylor EB, Rutter J. Mitochondrial quality control by the ubiquitin-proteasome system. *Biochem Soc Trans.* 2011;39:1509–1513.
63. Labbadia J, et al. Mitochondrial stress restores the heat shock response and prevents proteostasis collapse during aging. *Cell Rep.* 2017;21:1481–1494.
64. Schon EA, Manfredi G. Neuronal degeneration and mitochondrial dysfunction. *J Clin Invest.* 2003;111:303–312.

65. Ruan L, et al. Cytosolic proteostasis through importing of misfolded proteins into mitochondria. *Nature.* 2017;543:443–446.
66. Bjedov I, et al. Mechanisms of life span extension by rapamycin in the fruit fly *Drosophila melanogaster. Cell Metab.* 2010;11:35–46.
67. Rubinsztein DC, Marino G, Kroemer G. Autophagy and aging. *Cell.* 2011;146:682–695.
68. Madeo F, Zimmermann A, Maiuri MC, Kroemer G. Essential role for autophagy in life span extension. *J Clin Invest.* 2015;125:85–93.
69. Farre JC, Subramani S. Mechanistic insights into selective autophagy pathways: lessons from yeast. *Nat Rev Mol Cell Biol.* 2016;17:537–552.
70. Pickrell AM, Youle RJ. The roles of PINK1, parkin, and mitochondrial fidelity in Parkinson's disease. *Neuron.* 2015;85:257–273.
71. Sarraf SA, et al. Landscape of the PARKIN-dependent ubiquitylome in response to mitochondrial depolarization. *Nature.* 2013;496:372–376.
72. Ho TT, et al. Autophagy maintains the metabolism and function of young and old stem cells. *Nature.* 2017;543:205–210.
73. Soubannier V, et al. A vesicular transport pathway shuttles cargo from mitochondria to lysosomes. *Curr Biol.* 2012;22:135–141.
74. Luo M, Zhao X, Song Y, Cheng H, Zhou R. Nuclear autophagy: an evolutionarily conserved mechanism of nuclear degradation in the cytoplasm. *Autophagy.* 2016;12:1973–1983.
75. Dou Z, et al. Autophagy mediates degradation of nuclear lamina. *Nature.* 2015;527:105–109.
76. Hohn A, Grune T. Lipofuscin: formation, effects and role of macroautophagy. *Redox Biol.* 2013;1:140–144.
77. Folick A, et al. Aging. Lysosomal signaling molecules regulate longevity in *Caenorhabditis elegans. Science.* 2015;347:83–86.
78. de Cabo R, Carmona-Gutierrez D, Bernier M, Hall MN, Madeo F. The search for antiaging interventions: from elixirs to fasting regimens. *Cell.* 2014;157:1515–1526.
79. Heintz C, Mair W. You are what you host: microbiome modulation of the aging process. *Cell.* 2014;156:408–411.
80. Johnson SC, Rabinovitch PS, Kaeberlein M. mTOR is a key modulator of ageing and age-related disease. *Nature.* 2013;493:338–345.
81. Junnila RK, List EO, Berryman DE, Murrey JW, Kopchick JJ. The GH/IGF-1 axis in ageing and longevity. *Nat Rev Endocrinol.* 2013;9:366–376.
82. Salminen A, Kaarniranta K. AMP-activated protein kinase (AMPK) controls the aging process via an integrated signaling network. *Ageing Res Rev.* 2012;11:230–241.
83. Guarente L. Sirtuins in aging and disease. *Cold Spring Harb Symp Quant Biol.* 2007;72:483–488.
84. Martins R, Lithgow GJ, Link W. Long live FOXO: unraveling the role of FOXO proteins in aging and longevity. *Aging Cell.* 2016;15:196–207.
85. Guo S, et al. Phosphorylation of serine 256 by protein kinase B disrupts transactivation by FKHR and mediates effects of insulin on insulin-like growth factor-binding protein-1 promoter activity through a conserved insulin response sequence. *J Biol Chem.* 1999;274:17184–17192.
86. Houtkooper RH, Williams RW, Auwerx J. Metabolic networks of longevity. *Cell.* 2010;142:9–14.
87. Mihaylova MM, Shaw RJ. The AMPK signalling pathway coordinates cell growth, autophagy and metabolism. *Nat Cell Biol.* 2011;13:1016–1023.
88. Fernandez-Marcos PJ, Auwerx J. Regulation of PGC-1alpha, a nodal regulator of mitochondrial biogenesis. *Am J Clin Nutr.* 2011;93:884S–S890.
89. Price NL, et al. SIRT1 is required for AMPK activation and the beneficial effects of resveratrol on mitochondrial function. *Cell Metab.* 2012;15:675–690.

90. Kaur J, Debnath J. Autophagy at the crossroads of catabolism and anabolism. *Nat Rev Mol Cell Biol.* 2015;16:461–472.
91. Schaffer BE, et al. Identification of AMPK hosphorylation sites reveals a network of proteins involved in cell invasion and facilitates large-scale substrate prediction. *Cell Metab.* 2015;22:907 921.
92. Singh R, Cuervo AM. Autophagy in the cellular energetic balance. *Cell Metab.* 2011;13:495–504.
93. Meng S, et al. Metformin activates AMP-activated protein kinase by promoting formation of the alphabetagamma heterotrimeric complex. *J Biol Chem.* 2015;290:3793–3802.
94. Mair W, et al. Lifespan extension induced by AMPK and calcineurin is mediated by CRTC-1 and CREB. *Nature.* 2011;470:404–408.
95. Mostoslavsky R, et al. Genomic instability and aging-like phenotype in the absence of mammalian SIRT6. *Cell.* 2006;124:315–329.
96. Canto C, et al. AMPK regulates energy expenditure by modulating NAD+ metabolism and SIRT1 activity. *Nature.* 2009;458:1056–1060.
97. Mohrin M, et al. Stem cell aging. A mitochondrial UPR-mediated metabolic checkpoint regulates hematopoietic stem cell aging. *Science.* 2015;347:1374–1377.
98. Carrero D, Soria-Valles C, Lopez-Otin C. Hallmarks of progeroid syndromes: lessons from mice and reprogrammed cells. *Dis Model Mech.* 2016;9:719–735.
99. Dobrzynska A, Gonzalo S, Shanahan C, Askjaer P. The nuclear lamina in health and disease. *Nucleus.* 2016;7:233–248.
100. Kubben N, et al. Repression of the antioxidant NRF2 pathway in premature aging. *Cell.* 2016;165:1361–1374.
101. Cao K, et al. Rapamycin reverses cellular phenotypes and enhances mutant protein clearance in Hutchinson-Gilford progeria syndrome cells. *Sci Transl Med.* 2011;3:89ra58.
102. Krishnamurthy J, et al. Ink4a/Arf expression is a biomarker of aging. *J Clin Invest.* 2004;114:1299–1307.
103. Jeck WR, Siebold AP, Sharpless NE. Review: a meta-analysis of GWAS and age-associated diseases. *Aging Cell.* 2012;11:727–731.
104. Dolivo D, Hernandez S, Dominko T. Cellular lifespan and senescence: a complex balance between multiple cellular pathways. *Bioessays.* 2016;38(Suppl 1):S33–S44.
105. Sahin E, et al. Telomere dysfunction induces metabolic and mitochondrial compromise. *Nature.* 2011;470:359–365.
106. Park JY, et al. p53 improves aerobic exercise capacity and augments skeletal muscle mitochondrial DNA content. *Circ Res.* 2009;105:705–712.
107. Pal S, Tyler JK. Epigenetics and aging. *Sci Adv.* 2016;2:e1600584.
108. Bahar R, et al. Increased cell-to-cell variation in gene expression in ageing mouse heart. *Nature.* 2006;441:1011–1014.
109. Huh I, Zeng J, Park T, Yi SV. DNA methylation and transcriptional noise. *Epigenetics Chromatin.* 2013;6:9.
110. Wood JG, Helfand SL. Chromatin structure and transposable elements in organismal aging. *Front Genet.* 2013;4:274.
111. Sedivy JM, et al. Death by transposition—the enemy within? *Bioessays.* 2013;35:1035–1043.
112. Greer EL, et al. Transgenerational epigenetic inheritance of longevity in *Caenorhabditis elegans*. *Nature.* 2011;479:365–371.
113. Daxinger L, Whitelaw E. Transgenerational epigenetic inheritance: more questions than answers. *Genome Res.* 2010;20:1623–1628.
114. Petersen KF, Dufour S, Befroy D, Garcia R, Shulman GI. Impaired mitochondrial activity in the insulin-resistant offspring of patients with type 2 diabetes. *N Engl J Med.* 2004;350:664–671.

115. Zhang W, et al. Aging stem cells. A Werner syndrome stem cell model unveils heterochromatin alterations as a driver of human aging. *Science.* 2015;348:1160–1163.
116. Ashapkin VV, Kutueva LI, Vanyushin BF. Aging as an epigenetic phenomenon. *Curr Genomics.* 2017;18:385–407.
117. Taylor GC, Eskeland R, Hekimoglu-Balkan B, Pradeepa MM, Bickmore WA. H4K16 acetylation marks active genes and enhancers of embryonic stem cells, but does not alter chromatin compaction. *Genome Res.* 2013;23:2053–2065.
118. Brunet A, Rando TA. Interaction between epigenetic and metabolism in aging stem cells. *Curr Opin Cell Biol.* 2017;45:1–7.
119. Chandel NS, Jasper H, Ho TT, Passegue E. Metabolic regulation of stem cell function in tissue homeostasis and organismal ageing. *Nat Cell Biol.* 2016;18:823–832.
120. Gut P, Verdin E. The nexus of chromatin regulation and intermediary metabolism. *Nature.* 2013;502:489–498.
121. Su X, Wellen KE, Rabinowitz JD. Metabolic control of methylation and acetylation. *Curr Opin Chem Biol.* 2016;30:52–60.
122. Wu N, et al. Alpha-Ketoglutarate: physiological functions and applications. *Biomol Ther.* 2016;24:1–8.
123. Canto C, Menzies KJ, Auwerx J. NAD(+) metabolism and the control of energy homeostasis: a balancing act between mitochondria and the nucleus. *Cell Metab.* 2015;22:31–53.
124. Peleg S, Feller C, Ladurner AG, Imhof A. The metabolic impact on histone acetylation and transcription in ageing. *Trends Biochem Sci.* 2016;41:700–711.
125. Peleg S, et al. Life span extension by targeting a link between metabolism and histone acetylation in *Drosophila*. *EMBO Rep.* 2016;17:455–469.
126. Verdin E. NAD(+) in aging, metabolism, and neurodegeneration. *Science.* 2015;350:1208–1213.
127. Oldham WM, Clish CB, Yang Y, Loscalzo J. Hypoxia-mediated increases in L-2-hydroxyglutarate coordinate the metabolic response to reductive stress. *Cell Metab.* 2015;22:291–303.
128. Anso E, et al. The mitochondrial respiratory chain is essential for haematopoietic stem cell function. *Nat Cell Biol.* 2017;19:614–625.
129. Dang L, et al. Cancer-associated IDH1 mutations produce 2-hydroxyglutarate. *Nature.* 2009;462:739–744.
130. Lee IH, Finkel T. Metabolic regulation of the cell cycle. *Curr Opin Cell Biol.* 2013;25:724–729.
131. Wiley CD, et al. Mitochondrial dysfunction induces senescence with a distinct secretory phenotype. *Cell Metab.* 2016;23:303–314.
132. Merkwirth C, et al. Two conserved histone demethylases regulate mitochondrial stress-induced longevity. *Cell.* 2016;165:1209–1223.
133. Han S, et al. Mono-unsaturated fatty acids link H3K4me3 modifiers to *C. elegans* lifespan. *Nature.* 2017;544:185–190.

CHAPTER EIGHT

A Comparative Approach to Metabolic Aspects of Aging: Conserved Mechanisms and Effects of Calorie Restriction and Environment

Mary Ann Ottinger[1]

University of Houston, Houston, TX, United States

[1] Corresponding author. E-mail address: maotting@central.uh.edu

Contents

Abstract

Metabolic systems and the function of these systems are complex, involving biochemical pathways, endocrine, neuroendocrine systems, and physiological systems and interact with environmental conditions. Studies in animal models have been invaluable in gaining an understanding of the mechanisms involved in metabolic endocrine changes during normal aging and with conditions, such as diabetes and obesity. Together, these studies have revealed some conserved mechanisms and identified specific biomarkers of aging related to metabolic changes. Further, characterization of these mechanisms provides an opportunity to develop interventions and treatments

Progress in Molecular Biology and Translational Science, Volume 155
ISSN 1877-1173
http://dx.doi.org/10.1016/bs.pmbts.2017.11.004

for both humans and other vertebrates. This chapter will provide an overview of age-related changes in metabolism from studies in human populations and the perspective of information gained from comparative animal models. Detailed molecular mechanisms and endocrine pathways have already been discussed in other chapters of this volume. Finally, calorie restriction (CR) has shown consistent benefit to age-related disease incidence with effects that has been consistent across animal models These studies on the effects of CR enable further discernment of disease versus healthy aging processes.

1. INTRODUCTION

There is tremendous variation in the process of aging across individuals and species. Insights from studies on animal models have provided an understanding of both fundamental processes and how these processes become altered during of aging. Comparison of short- and long-lived species provides an excellent platform for comparison of biological processes; birds are going to be considered in this review as one example of a comparative model of aging. It is clear that metabolic changes are key drivers of age-related declining physiological functions, which are involved in virtually all physiological systems, cognitive function, and ultimately healthy aging and longevity. One of the fundamental characteristics of aging and metabolism in many individuals is the development of metabolic syndrome, in which slowing metabolisms is accompanied by increasing insensitivity in glucose modulation and energy utilization. Further, the characteristics of metabolic syndrome are appearing more frequently in younger individuals with the rising incidence of obesity. It is difficult to identify specific triggers that initiate age-related changes, as well as assess impacts of selected functional alterations. However, it is clear that inflammation is a key component and that metabolic disease and dysfunction is interrelated to inflammatory responses. One of the most ubiquitous conditions associated with aging is metabolic syndrome, which impact a majority of adults as they age. More often it is defined in the context of a loss of responsiveness of metabolic systems resulting in altered glucose metabolism, reduced efficiency in energy utilization, and reduced overall health. However, metabolic syndrome is also becoming a condition in younger individuals with the rising incidence of obesity. Other biomarkers, as well as progenitors of the age-related metabolic decline include increasing oxidative damage, altered endocrine, and neuroendocrine function, specifically involving growth hormone, thyroid,

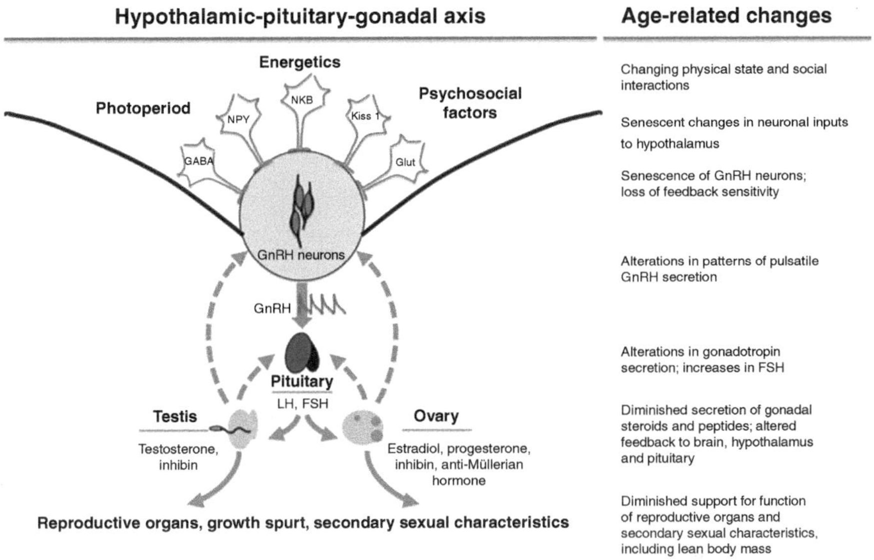

Fig. 1 The hypothalamic-pituitary-gonadal axis in birds and mammals and associated age-related changes. *FSH*, Follicle stimulating hormone; *GABA*, gamma-aminobutyric acid; *Glut*, glutamine; *GnRH*, gonadotropin-releasing hormone; *Kiss 1*, kisspeptin q; *LH*, luteinizing hormone; *NKB*, neuokinin B; *NPY*, neuropeptide Y. *From Ellison, PE, Ottinger, MA, Ellison PE, Ottinger MA. A comparative perspective on reproductive aging, reproductive cessation, post-reproductive life, and social behavior. In: Weinstein M, Lane MA, eds. Sociality, Hierarchy, Health Comparative Biodemography: A collection of papers. Committee on Population, Division of Behavioral and Social Sciences and Education, National Research Council of the National Academies, National Academy of Sciences; 2014:315–338.*[1]

reproductive, and adrenal systems. The timing and impact of the age-related decline in reproductive function may be exacerbated by metabolic dysfunction (Fig. 1).[1] Conversely, dietary factors and exercise may help ease associated symptoms in both males and females. Circadian rhythms in metabolic hormones and physiological function become muted with aging. As such, early incidence of obesity and the impacts of environmental factors become significant in the initiation and progression of age-related metabolic functional decline. Interventions, such as calorie restriction (CR) have been effective, promoting healthy aging but not resulting in increasing longevity in primates. An overview of metabolic syndrome and other age-related alterations in metabolism follows, with an overview of CR effects and findings from selected animal models.

2. METABOLISM, INFLAMMATION AND AGING

Inflammation and disease has become a central issue for the understanding and treatment of disease. Chronic stress has been linked to the development of type-2 diabetes and diet, metabolism, and inflammation intersect to trigger the onset and disease progression for atherosclerosis.[2] Stressors that begin earlier in life may disrupt homeostatic mechanisms with dietary and behavioral habits that leading to chronic disruption of metabolic neuroendocrine modulators and ultimately inflammatory disease (for review, see Ref. 3). This disturbed balance in metabolism and stressors resulting in reward-driven food intake and other coping mechanisms plays a strong role in today's society and can ultimately impact longevity and healthy aging. Promising studies show that dietary treatments can serve as an intervention approach for specific autoimmune diseases.[4]

3. METABOLIC SYNDROME AND OTHER BIOMARKERS OF METABOLIC AGING

Metabolic syndrome describes a constellation of disorders involving pre-diabetes, glucose intolerance, hypertension, obesity, and associated inflammatory disease and potentially cardiovascular disease. Other conditions often emerge in the overall unhealthy state of metabolic endocrine function and impacts on other physiological systems. Chronic inflammation accompanying cardiovascular disease, diabetes and other conditions have been linked with metabolic syndrome and production of cytokines by adipose tissue; more data are needed to understand the underlying mechanisms.[5–7] It is well recognized that adipose tissue is highly metabolically active producing the adipocytokines, adiponectin, and leptin.[8] These adipocytokines appear to relate to longevity and are clearly linked with the functional integrity of the metabolic system as an interactive entity (for review, see Ref. 8–10). This functional integrity involves many facets of endocrine function, including the somatotropic and glucose regulatory systems, which then impact longevity.[11] It is also clear that brain health and specifically cognitive function are greatly affected by unstable circulating blood glucose levels.[12] This relationship becomes more extreme in the case of Alzheimer's disease in which neuroinflammation, oxidative damage, and metabolic dysfunction converge to promote disease progression and in the relationship between diabetes and cancer.[10,13,14] The reproductive endocrine system is impacted by metabolic syndrome, with reduced circulating testosterone levels in aging males.[15] This observation is similar to those in other vertebrates in which obesity and metabolic dysfunction correlate with low circulating testosterone and elevated circulating estrogens, likely associated with steroid metabolism in adipose tissue. Recommendations from an international panel include diet emphasizing the Mediterranean-type diet, with CR if preferred.[16] There is also increasing attention to the role of the microbiome in the effective absorption and utilization of nutrient as influential in the process of normal aging and in disease states, such as Alzheimer's disease.[17] Further, lifestyle interventions, including CR and exercise, serve to modulate brain health during aging and with metabolic stressors.[18]

4. OXIDATIVE DAMAGE

Oxidative stress occurs throughout the lifespan and culminates with increased impacts from the damage due to free radical production coincident with aging. The oxidative stress theory of aging encapsulates the sequence of lifetime events, drawing from a rich literature, which includes investigations of the stressors, molecular mechanisms involved in the adverse effects from free radical production associated with the formation of reactive oxygen species (ROS). It is clear that there is an increased pathology from exposure to oxidative stress that can affect all physiological systems.[19] However, it is not clear if this affects longevity. Based on findings in recent studies, it has been hypothesized that the cumulative effect of oxidative damage must be considered for assessing lifetime effects, specifically relative to the number of cells that become impaired and senescent due to oxidative damage become nonfunctional areas in an organ or tissue.[20] This is an interesting concept—it begins to explain some of the variability in sensitivity to stressors, such as oxidative damage, as well as the occasional lack of discernable impact specifically attributable to oxidative damage. Moreover, there are potential connections that need to be examined between overall energy levels during aging and mitochondrial function at tissue and organ levels as a critical facet of optimal physiological function.

Comparative studies provide supporting evidence for the importance of ROS production and the relationship between longevity and resistance or compensation to oxidative stress. Conserved mechanisms have been shown across vertebrate and invertebrate species, with long-lived species showing less oxidative damage than short-lived species.[21] Further, an accelerating aging phenotype has been observed in mice deficient in antioxidant enzymes Cu/Zn-superoxide dismutase.[20] Short-lived birds show lower resistance to damage from oxidative damage compared to long-lived birds.[22–25] However, an evidence showing no difference in oxidative damage and associated cellular metabolic processes have been reported by Montgomery *et al.*[26,27] The contradictory an evidence may be due in part to differential cellular responses to oxidative stress accompanying atherosclerosis, quail microvascular endothelial cells show greater resilience by production of heme oxygenase as protection against oxidized lipids.[28] Taken together, it is clear that these processes are complex and there is intriguing evidence for a differential and potentially advantageous mechanisms, especially in long-lived birds.

5. CIRCADIAN RHYTHMS

The role of circadian rhythms has been recognized as both a biomarker of age-related declining rhythms, including sleep wake cycles and feeding cycles.[29] Circadian rhythms in hormones, neural system responses, sleep cycles, and other physiological functions become less pronounced and diminished in amplitude and sometimes frequency during aging. There is evidence for effects of age on clock gene expression in the rhesus macaque pituitary gland.[30] There is a compelling evidence that CR and circadian clocks interact; specifically by modulation of nutrient signaling pathways that then impact metabolic pathways.[30a] These signaling pathways could provide an avenue for potential treatments during aging to restore muted circadian rhythms and maintain sleep-wake cycles. Moreover, molecular clocks directing circadian rhythms modulate bioenergetics at all levels and there is a growing evidence for patterns of mitochondrial respiration relate with production and decay of amyloid-beta.[31]

6. CALORIE RESTRICTION

CR has been one of the few interventions demonstrating beneficial effects across a number of species.[31a–33] Some of the effects of CR appear to directly relate to reduced or resistance to oxidative damage, diabetes, and other physiological conditions associated with aging and potential disease. Merry and Holehan[34] documented effects of CR on sexual maturation, and subsequent studies showed that CR could reduce reproduction or in some cases may extend reproductive lifespan.[35–37] The degree of CR has proven critical as different species show varied sensitivity. Reproduction has also proven to be an interesting responsive measure because males and females appear to respond differentially overtime. Moreover, it has been documented in humans and other species that harsh conditions, which in this case is severe or in some species even moderate CR above 25% can result in diminished reproductive function. There is a rich literature for the extensive work that has been conducted using transgenic mouse models to understand metabolic endocrine pathways as affected by CR, as well as the mechanisms associated with impacts on longevity (for review, see Ref. 38).

Domestic poultry provides an interesting set of data because CR has been routinely used to optimizing productivity and health.[39] As in mammals, CR during maturation slows the onset of puberty. Ultimately, CR also enhances reproductive function and productivity in poultry, with more chicks hatched and extended reproductive lifespan.[40]

Data from primates have shown benefit from moderate CR, in terms of lipid metabolism and delayed age-related diabetes.[40a–42] As in rodents and other phyla, including birds, there is a great deal of interest in the effects of CR on oxidative damage, free radical production, and thyroid system function during aging.[41,42a]

Data collected in male rhesus monkeys document similar responses to CR, in that initiation of 30% CR during puberty resulted in a transitory delay in sexual maturation, which was also reflected in lower androgen levels.[42b]

In humans, there has been a "comprehensive assessment of the long-term effects of reducing intake of energy (CALERIE)" has shown positive results from moderate CR.[43] Interestingly, intermittent CR has counteracted on diet-induced obesity in aging male mice.[43a] In addition to CR effects on reducing oxidative stress, there is also an evidence for some level of oxidative stress being important to physiological function.[44] It has been thought that moderate CR is perceived by the organism as mild stress, inducing a hormetic response, which stimulates physiological systems to perform more efficiently.

7. INSIGHTS FROM SHORT- AND LONG-LIVED BIRDS AS MODELS OF AGING

There are thousands of species of birds spanning the globe in a variety of habitats. Further, birds vary greatly in lifespan, habitat, adaptations to climate, and resources and have specialized characteristics that allow long distance migration and survival under extreme conditions.[45] Birds also have higher metabolic rate and body temperature, with unique bone and air sac structure to enhance flight capability. As such, it would be predicted that the higher body temperature and metabolic rate would result in a shorter lifespan. Many birds, including several species of hummingbirds, parrots, and seabirds contradict this prediction and have remarkably long lifespans.[45a–51] The question of how this is possible, given the usual predictors of accelerating aging, such as oxidative damage, which is surprisingly reduced compared to short-lived birds. Similarly, other diseases occur, but much later in the lifespan including cancer and endocrine system decline. As such, the aging of neural systems modulating reproductive endocrine and metabolic systems reflect lifespan related functional impacts as the aging process progresses.

Studies on the physiology of birds, including embryonic development have come in large part from domestic species; however, there is a growing literature from wild birds studied in the field and laboratory. Further, field studies provide tremendous insight into the adaptations of wild birds to environmental conditions and stressors. Studies in domestic species often focus on age-related decline in reproduction resulting in reduced egg production and fertility.[52–56] Alterations in metabolic function are reflected in thinning eggshell, which occurs with decreasing egg production.[57] Coincident with the declining egg production, other indicators of age-related decline begin to emerge, including alterations in neuroendocrine response, shell quality decline, and increasing evidence of metabolic system changes.[52–54,56] Osteoporosis often presents severe problems for aging domestic birds, especially in egg-laying strains that may produce 300+ eggs/year because the bones serve as a repository for calcium and the other minerals needed for egg production; the shell also becomes progressively thinner.[56a] As in humans, exercise by providing perches improves bone and muscle health (Hester *et al.*, 2003). Short-lived male birds also show evidence of aging, with decreased sperm production, increased sperm abnormalities, and reduced circulating hormones, changes in neuroendocrine systems accompanied by diminished male sexual behavior.[53,58] Interestingly, aging male and female birds have a higher incidence of ovarian and testicular cancer.

Biomarkers of aging have been examined in birds to determine if there are conserved mechanisms signaling the initial phases of the aging process. Comparison of putative biomarkers in short- and long-lived birds point to changes in specific systems, including altered immune system function that may occur in tandem with reproductive endocrine decline.[48,51] Further, the link between environmental stressors and healthy aging is critical, affecting both health and longevity.[59] Telomere length, which has longer DNA sequences in birds compared to mammals were not related to maximum life span.[60] However,[61] demonstrated declining telomere length in African Grey parrots, especially those that were socially isolated. As such, stressors are critical to the health of cellular metabolic systems.[61a] Therefore, stress and potentially oxidative damage as discussed earlier influences cellular function and the efficiency of the mitochondrial energy producing system, both of which decline during aging. As such, short- and long-lived birds provide unique comparable models of aging; they also show extended lifespan by living in cooperative societal social structures.[62] Further, environmental conditions in early life may impact long-term survival.[63] Neurotransmitter and neuroendocrine systems are a clear and key component of aging, affecting endocrine and behavioral responses.[46,53,64] As such, there are parallel processes appear to be occurring during aging across vertebrates that impact similar physiological systems and contribute to decline in metabolic endocrine function and in some cases the onset of disease.

8. EFFECTS OF ENVIRONMENTAL ENDOCRINE ACTIVE CHEMICALS AND METABOLIC AGING

There has been a dramatic increase in the use chemicals in industrial, agricultural, and residential treatments, resulting in a diversity of both native and manufactured chemicals utilized across a range of applications. While technology has greatly benefited society, there have been associated unintended adverse effects associated with many of these environmental contaminants. However, assessing the potential for adverse effects due to the vast range of contaminants in our environment has been difficult due to the diversity of actions and potencies, bioavailability, life cycle of compounds, and exposure scenarios. Further, assessing the potential of associated health impacts is exceedingly complex with consideration of short-term toxicological versus long-term cumulative adverse effects on human, domestic species, and wildlife health. Beyond frank toxicity, endocrine disrupting chemicals (EDCs) have emerged and present a more difficult issue in terms of ascertaining risk because EDCs are often nonlethal in field relevant exposures while exerting hormone-like effects. These effects may interfere with a number of endocrine systems and processes, not the least of which is reproductive or metabolic endocrine function. As such, EDCs have received attention as associated with human health issues, specifically obesity and reproductive health. While domestic species may experience similar effects due to their proximity to humans, wildlife are exposed to a myriad of environmental chemicals, including metals, pesticides, fungicides, and industrial chemicals. Both provide additional insights into short- and long-term effects of EDCs. Interestingly, many plant produce hormones that also affect vertebrate species and thereby contribute to the combined effects of exposure to environmental chemicals, especially endocrine disruptors. Not only is there a vast array of compounds that are suspected EDCs, but also the mechanisms of action for these compounds are extremely diverse, both in their targets and actions. Moreover, differences in response and sensitivity to specific compounds do vary across individuals, species, and phyla. This is especially relevant when considering variations in reproductive processes and strategies, which also translate into species differences in developmental patterns. Often, the phase of the life cycle that is most sensitive to effects of EDC exposure will also vary with the timing of these ontogenic patterns and events. As a result, it becomes extremely complex to develop clear indices of EDC impact that apply across a number of species, even in one phylum. Considerable historical data exist which are reviewed in a number of publications.[64a–70]

Attempts to assess the action(s) and potency of EDCs have incorporated several approaches, including: (1) mapping the usage high volume use chemicals in agriculture, residential, or industry; (2) measuring field concentrations of selected compounds and linking these data to field surveys of wildlife populations; (3) performing bioassays and structure-function models to assess action and potency; (4) ascertaining geographic "hot spots" for the incidence of adverse effects on human health and determining possible environmental correlate(s); and (5) conducting controlled laboratory studies to characterize potential endocrine actions of specific chemicals or mixtures of chemicals that mimic field observations. This review presents an overview of some high volume pesticides with consideration of both toxicological and endocrine disrupting actions. Many EDCs are nonlethal, and generally categorized into broad classes, identified on the basis of their physiological actions. Polychlorinated biphenyls will be discussed in greater detail as an example of a family of compounds that have been heavily used across a range of applications, yet are long-lived and linked to adverse effects in human and wildlife health. Further, considerable data have been gathered for the PCBs, which remain ubiquitous in the environment and have both toxicological and endocrine disrupting actions.

8.1 High Volume Chemical Usage and Assessing Potential Endocrine Disruption

There has been extensive study of the fate and effects of chemicals in the environment, especially related to impacts on human health. Particular attention has been directed toward food safety, and more specifically on the residues of pesticides and fungicides on fruits and vegetables. A comprehensive list of pesticide residues may be found for fruits and vegetables on the USDA Pesticide Data Program (https://www.ams.usda.gov/datasets/pdp) and will not be discussed in detail here. With agricultural application, there is also the risk of runoff of chemicals into waterways. As such, regions of high volume application of pesticides are more likely to have contaminated waterways and in some cases, groundwater contamination. Ryberg and Gilliom[71] found that pesticides used in agriculture, such as alachlor, atrazine, and carbofuran showed high correlation between use and detected concentrations; however, chemicals used in both agricultural and nonagricultural showed more correlation to when urban use was added to the model. Accordingly, it is important to consider both use and subsequent fate and life cycle of EDCs. Mapping of the use of selected pesticides provides insight into the relationship between agricultural application and subsequent detection in waterways and groundwater. As seen in Fig. 1A, in 2014 the herbicide atrazine has been used in high volumes across the United States with highest concentrations used in the midsection of the country on cornfields. Fig. 1B shows estimates of atrazine and deethylatrazine concentrations in the groundwater associated with agricultural areas. Data are mixed regarding any effects of atrazine and its metabolite, deethylatrazine on vertebrate species. Some studies suggest that atrazine may interact with developmental processes; however, the data have not conclusively shown a causal relationship (https://www.epa.gov/ingredients-used-pesticide-products/atrazine-background-and-updates). As such, the use of atrazine will continue unless there are further data indicating human or environmental risks.

Diazinon was heavily used as an insecticide from the 1950s until it was banned for use with the exception agricultural use (Fig. 2). It is an organophospate insecticide that targets the central neurotransmitter acetylcholinesterase (AChE) system. As with other organophosphate compounds, there are severe toxicity risks impacting neural systems, which exert both

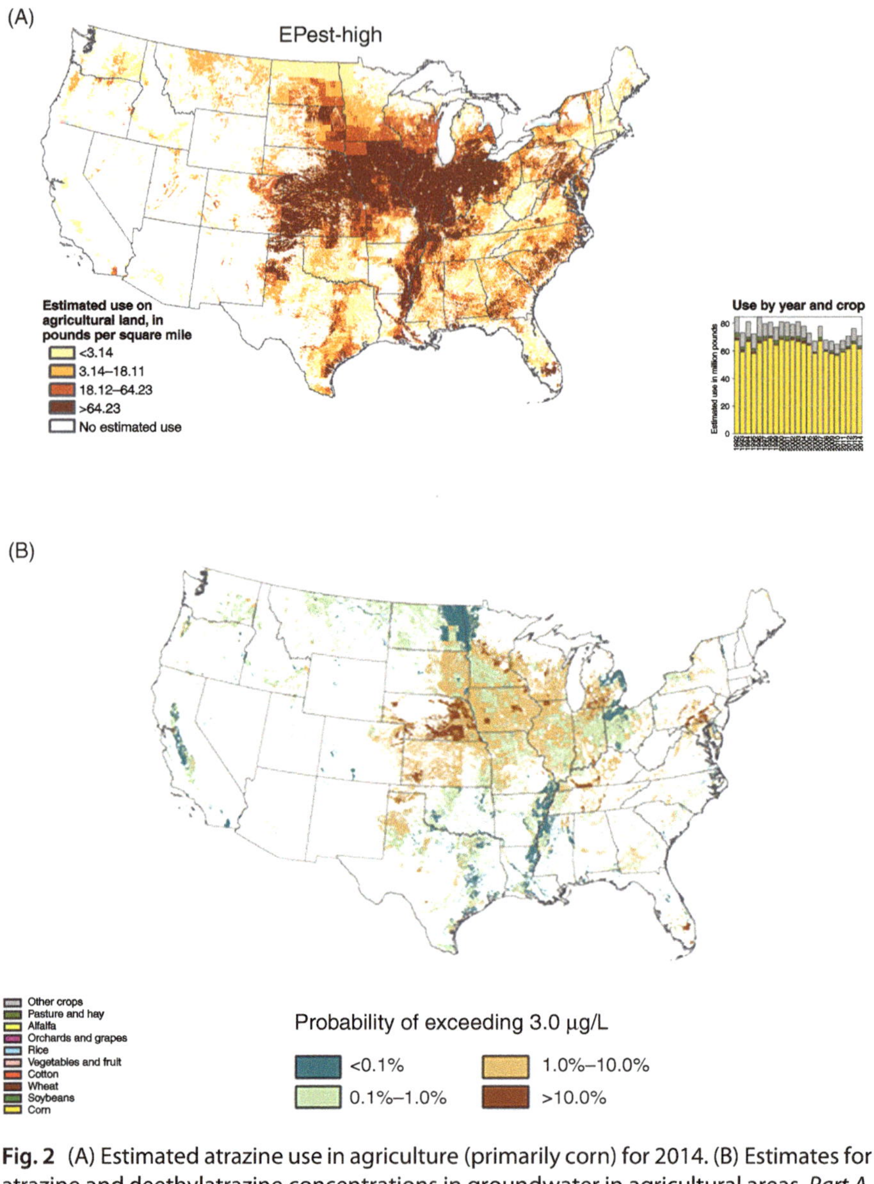

Fig. 2 (A) Estimated atrazine use in agriculture (primarily corn) for 2014. (B) Estimates for atrazine and deethylatrazine concentrations in groundwater in agricultural areas. *Part A, from Pesticide National Synthesis Project; USGS https://water.usgs.gov/nawqa/pnsp/usage/maps/show_map.php?year=2014&map=ATRAZINE&hilo=H&disp=Atrazine; part B, Pesticide National Synthesis Project; map from https://water.usgs.gov/nawqa/pnsp/features/feature.php; Stackelberg PE, Barbash JE, Gilliom RJ, Stone WW, Wolock DM. Regression models for estimating concentrations of atrazine plus deethylatrazine in shallow groundwater in agricultural areas of the United States. J Environ Qual. 2012; 41(2):479–494.*

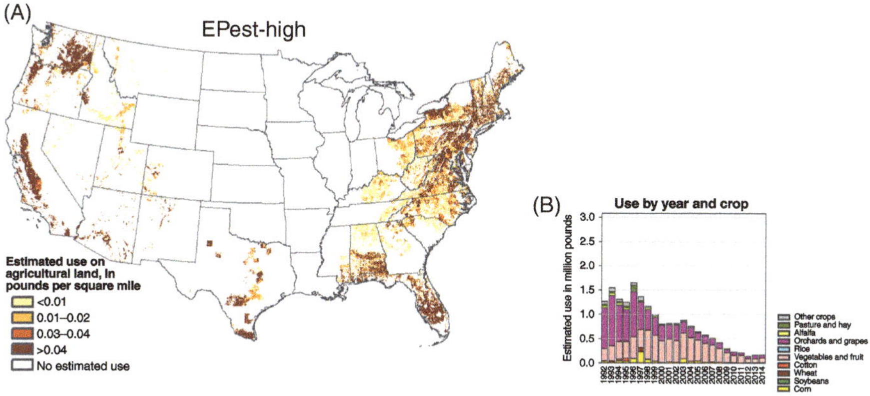

Fig. 3 (A) Map showing estimated use of diazinon in agriculture for 2014. (B) Chart showing use across agricultural crops. *Part A, from https://water.usgs.gov/nawqa/pnsp/usage/maps/show_map.php?year=2014&map=DIAZINON&hilo=H&disp=Diazinon.*

toxicological and in low doses impact endocrine and other physiological systems (see https://www.epa.gov/endangered-species/biological-evaluation-chapters-diazinon-esa-assessment for Biological Evaluation Chapters for Diazinon ESA Assessment).

Chlorpyrifos is a high volume pesticide used for agricultural and nonagricultural applications (Fig. 3). It has been in use since 1965; there is concern about occupational exposure because cholinesterase inhibition can occur at higher doses. The Environmental Protection Agency (EPA) has put limitation and restrictions on the use of chlorpyrifos. The EPA conducted an in depth review of potential adverse effects of chlorpyrifos on neurodevelopmental processes associated with residues on fruit and vegetables (see https://www.epa.gov/endangered-species/biological-evaluation-chapters-chlorpyrifos-esa-assessment for detailed reports).

8.2 EDCs and Obesity

The link between EDCs and obesity was documented some time ago and these estrogenic compounds, termed obesogens were shown to impact metabolic function, especially with exposure during development.[72] Similarly, obesogens are likely candidates in the disruption of adipose tissue function, again with developmental exposure potentially impacting lifetime metabolic function (for review, see Ref. 73,74). Recent data from humans support the potential risk from EDCs on the pathogenesis fatty liver disease, type 2 diabetes, and an associated form of metabolic syndrome.[74a,75]

Epidemiological studies in humans indicate that the rising incidence of chronic metabolic diseases, with increasing obesity and diabetes may relate to early EDC effects and in particular result in the permanent programming of metabolic endocrine systems.[76] A potential mechanism is via modulation of hormone receptor function, which in turn could have implications for epigenetic outcomes.[77] To this point, there is evidence for transgenerational effects of obesogens, affecting the health and metabolic endocrine function in offspring associated with exposure in the parental generation, meaning that the consequences of exposure persist over multiple generations.[78] As such, EDCs can have multiple targets that will exert short- and long-term effects on the organism's physiological function.

9. CONCLUSIONS

Understanding the complexities of normal aging with an overlay of conditions and diseases that arise during the process of aging is challenging. It is critical to consider all levels of function from molecular and subcellular responses to those of physiological systems and the entire organism. The influence and modulation of external factors is critical with a few of these factors being diet, nutrition, genetics, epigenetics, and a myriad of environmental chemicals. Critical influences for individuals and populations are social structure, resources, and lifestyle. Taken together, it is important to understand the interrelationships and interactions of these entities and formulate education and interventions to promote healthy aging.

REFERENCES

1. Ellison, PE, Ottinger, MA,Ellison PE, Ottinger MA. A comparative perspective on reproductive aging, reproductive cessation, post-reproductive life, and social behavior. In: Weinstein M, Lane MA, eds. *Sociality, Hierarchy, Health Comparative Biodemography: A collection of papers*. Committee on Population, Division of Behavioral and Social Sciences and Education, National Research Council of the National Academies, National Academy of Sciences; 2014:315–338
2. Ryan KK. Stress and metabolic disease. In: Weinstein M, Lane MA, eds. *Sociality, Hierarchy, Health Comparative Biodemography: A collection of papers*. Committee on Population, Division of Behavioral and Social Sciences and Education, National Research Council of the National Academies, National Academy of Sciences; 2014:247–268
3. Ulrich-Lai YM, Ryan KK. Neuroendocrine circuits governing energy balance and stress regulation: functional overlap and therapeutic implications. *Cell Metab*. 2014;19 (6):910–925.
4. Choi IY, Lee C, Longo VD. Nutrition and fasting mimicking diets in the prevention and treatment of autoimmune diseases and immunosenescence. *Mol Cell Endocrinol*. 2017;455:4–12, S0303-7207(17)30055-2.

5. Finch CE, Morgan TE, Longo VD, de Magalhaes JP. Cell resilience in species life spans: a link to inflammation? *Aging Cell.* 2010;9(4):519–526.
6. Lopez-Candales A, Hernandez Burgos PM, Hernandez-Suarez DF, Harris D. Linking chronic inflammation with cardiovascular disease: from normal aging to the metabolic syndrome. *J Nat Sci.* 2017;3(4). e341.
7. Stout MB, Justice JN, Nicklas BJ, Kirkland JL. Physiological aging: links among adipose tissue dysfunction, diabetes, and frailty. *Physiology.* 2017;32(1):9–19.
8. Gulcelik NE, Halil M, Ariogul S, Usman A. Adipocytokines and aging: adiponectin and leptin. *Minerva Endocrinol.* 2013;38(2):203–210.
9. Hamrick MW. Role of cytokine-like hormone leptin in muscle-bone crosstalk with aging. *J Bone Metab.* 2017;24(1):1–8.
10. Kang S, Lee YH, Lee JE. Metabolism-centric overview of the pathogenesis of Alzheimer's disease. *Yonsei Med J.* 2017;58(3):479–488.
11. Bartke A. Growth hormone, insulin and aging: the benefits of endocrine defects. *Exp Gerontol.* 2011 Feb-Mar;46(2–3):108–111.
12. Sunram-Lea SI, Owen L. The impact of diet-based glycaemic response and glucose regulation on cognition: evidence across the lifespan. *Proc Nutr Soc.* 2017;76(4):466–477.
13. Rojas-Gutierrez E, Munoz-Arenas G, Trevino S, Espinosa B, Chavez R, Rojas K, Flores G, Diaz A, Guevara J. Alzheimer's disease and metabolic syndrome: a link from oxidative stress and inflammation to neurodegeneration. *Synapse.* 2017;71:e21990.
14. Scappaticcio L, Maiorina MI, Bellastella G, Giugliano D, Esposito K. Insights into the relationships between diabetes, prediabetes, and cancer. *Endocrine.* 2017;56(2):231–239.
15. Blava R, Blava P, Rhoden L, Rhoden EL. Low testosterone levels and metabolic syndrome in aging male. *Curr Pharm Des.* 2017. doi: 10.2174/1381612823666170503150955.
16. Perez-Martinez P, Mikhailidis DP, Athyros VG, Bullo M, Couture P, covas MI, de Koning L, Delgado-Lista J, Diaz-Lopez A, Drevon CA, Estruch R, Esposito K, Fito M, Garaulet M, Giugliano D, Garcia-Rios A, Katsiki N, Kolovou G, Lamarche B, Maiorino MI, Mena-Sanchez G, Munoz-Garach A, Nikolic D, Ordovas JM, Perez-Jimenez F, Rizzo M, Salas-Salvado J, Schroder H, Tinahones FJ, de la Torre R, van Ommen B, Wopereis S, Ros E, Lopez-Miranda J. Lifestyle recommendations for the prevention and management of metabolic syndrome: an international panel recommendation. *Nutr Rev.* 2017;75(5):307–326.
17. Jiang C, Li G, Huang P, Liu Z, Zhao B. The gut microbiota and Alzheimer's disease. *J Alzheimers Dis.* 2017;58(1):1–15.
18. Camandola A, Mattson MP. Brain metabolism in health, aging, and neurodegeneration. *EMBO J.* 2017;36(11):1474–1492.
19. Zhang Y, Ikeno U, Wenbo Qi. Chaudhuri A, Li Y, Bokov A, Thorpe SR, Baynes JW, Epstein C, Richardson A, Van Remmen H. Mice deficient in both Mn superoxide dismutase and glutathione peroxidase-1 have increased oxidative damage and a greater incidence of pathology but no reduction in longevity. *J Gerontol Biol Sci Med Sci.* 2009;64 (12):1212–1220.
20. Zhang Y, Unnikrishnan A, Deepa SS, Liu Y, Ikeno Y, Sosnowska D, Van Remmen H, Richardson A. A new role for oxidative stress in aging: the accelerating aging phenotype in Sod1+ mice is correlated to increased cellular senescence. *Redox Biol.* 2017;11:30–37.
21. Pérez VI, Bokov A, Van Remmen H, Mele J, Ran Q, Ikeno Y, Richardson A. Is the oxidative stress theory of aging dead? *Biochim Biophys Acta.* 2009;1790(10):1005–1014.
22. Ogburn CE, Austad SN, Holmes DJ, Kiklevich JV, Gollahon K, Rabinovitch P, Martin GM. Cultured renal epithelial cells from birds and mice: enhanced resistance of avian cells to oxidative stress and DNA damage. *J Gerontol.* 1998;53A:B287–B292.
23. Ogburn CE, Martin GM, Ottinger MA, Holmes DJ, Carlberg K, Austad SN. Exceptional cellular resistance to oxidative damage in long-lived birds requires active gene expression. *J Gerontol.* 2001;11:B468–B474.

24. Herrero A, Barja G. 8-oxo-deoxyguanosine levels in heart and brain mitochondrial and nuclear DNA of two mammals and three birds in relation to their different rates of aging. *Aging*. 1999;11(5):294–300.
25. Pamplona R, Portero-Otín M, Riba D, Ledo F, Gredilla R, Herrero A, Barja G. Heart fatty acid unsaturation and lipid peroxidation, and aging rate, are lower in the canary and the parakeet than in the mouse. *Aging*. 1999;11:44–49.
26. Montgomery MK, Hulbert AJ, Buttemer WA. Does the oxidative theory of aging explain longevity differences in birds? I. Mitochondrial ROS production. *Exp Gerontol*. 2012;47(3):203–210.
27. Montgomery MK, Hulbert AJ, Buttemer WA. Metabolic rate and membrane fatty acid composition in birds: a comparison between long-living parrots and short-living fowl. *J Comp Physiol B*. 2012;182(1):127–137.
28. Hoekstra KA, Velleman SG. Brain microvascular and intracranial artery resistance to atherosclerosis is associated with heme oxygenase and ferritin in Japanese quail. *Mol Cell Biochem*. 2008;307(1–2):1–12.
29. Copinschi G, Caufriez A. Sleep and hormonal changes in aging. *Endocrinol Metab Clin North Am*. 2013;42(2):371–389.
30. Sitzmann BD, Lemos DR, Ottinger MA, Urbanski HF. Effects of age on clock gene expression in the rhesus macque pituitary gland. *Neurobiol Aging*. 2010;31(4):696–705.
30a. Chaudhari A, Gupta R, Makwana K, Kondratov R. Circadian clocks, diets and aging. *Nutr Healthy Aging*. 2017;4(2):101–112. doi: 10.3233/NHA-160006.
31. Schmitt K, Grimm A, Eckert A. Amyloid-β-induced changes in molecular clock properties and cellular bioenergetics. *Front Neurosci*. 2017;11:124.
31a. Lane MA, Mattson J, Ingram DK, Roth GS. Caloric restriction and aging in primates: Relevance to humans and possible CR mimetics. *Microsc Res Tech*. 2002;15(4):335–338.
32. Weindruch R, Walford RL. Dietary restriction in mice beginning at one year of age: effect on life span and spontaneous cancer incidence. *Science*. 1982;215:1415–1418.
33. Masoro EJ. Subfield history: calorie restriction, slowing aging, and extending life. *Sci Aging Knowl Environ*. 2003;2003(8):1–7.
34. Merry BJ, Holehan AM. Onset of puberty and duration of fertility in rats fed a restricted diet. *J Reprod Fertil*. 1979;57:253–259.
35. Nelson JF, Gosden RG, Felicio LS. Effect of dietary restriction on estrous cyclicity and follicular reserves in aging C57BL/6J mice. *Biol Reprod*. 1985;32:515–522.
36. Nelson JF, Karelus K, Bergman MD, Felicio LS. Neuroendocrine involvement in aging: evidence from studies of reproductive aging and caloric restriction. *Neurobiol Aging*. 1995;16(5):837–843.
37. McShane TM, Wise PM. Life-long moderate caloric restriction prolongs reproductive life span in rats without interrupting estrous cyclicity: effects on the gonadotropin-releasing hormone/luteinizing hormone axis. *Biol Reprod*. 1996;54:70–75.
38. Azzu V, Valencak TG. Energy metabolism and ageing in the mouse: a mini-review. *Gerontology*. 2017;63(4):327–336.
39. Miles SA, Leeson S. Effects of feed restriction during the rearing period and age at photostimulation on the reproductive performance of turkey hens. *Poult Sci*. 1990;69(9):1522–1528.
40. Renema RA, Robinson FE, Melnychuk VL, Hardin RT, Bagley LG, Emmerson DA, Blackman JR. The use of feed restriction for improving reproductive traits in male-line large white turkey hens. 1. Growth and carcass characteristics. *Poult Sci*. 1994;73(11):1724–1738.
40a. Colman RJ, Kemnitz JW, Lane MA, Abbott DH, Binkley N. Skeletal effects of aging and menopausal status in female rhesus macaques. *J Clin Endocrinol Metab*. 1999;84(11):4144–4148.

41. Lane MA, Black A, Handy A, Tilmont EM, Ingram DK, Roth GS. Caloric restriction in primates. *Ann N Y Acad Sci.* 2001;928:287–295.
42. Roth GS, Mattison JA, Ottinger MA, Chachich ME, Lane MA, Ingram DK. Aging in rhesus monkeys: relevance to human health interventions. *Science.* 2004;305: 1423–1426.
42a. Wei YH, Lee HC. Oxidative stress, mitochondrial DNA mutation, and impairment of antioxidant enzymes in aging. *Exp Biol Med (Maywood).* 2002;227(9):671–682.
42b. Urbanski HF, Downs JL, Garyfallou VT, Mattison JA, Lane MA, Roth GS, Ingram DK. Effect of caloric restriction on the 24-hour plasma DHEAS and cortisol profiles of young and old male rhesus macaques. *Ann N Y Acad Sci.* 2004;1019:443–447.
43. Das SK, Balasubramanian P, Weerasekara YK. Nutritional modulation of human aging: the calorie restriction paradigm. *Mol Cell Endocrinol.* 2017;455:148–157.
43a. Rusli F, Lute C, Boekschoten MV, van Diik M, van Norren K, Menke AL, Müller M, Steegenga WT. Intermittent calorie restriction largely counteracts the adverse health effects of a moderate-fat diet in aging C57BL/6J mice. *Mol Nutr Food Res.* 2017;61(5). doi: 10.1002/mnfr.201600677. Epub 2017 Mar 21.
44. Luo H, Chiang HH, Louw M, Susanto A, Chen D. Nutrient sensing and the oxidative stress response. *Trends Endocrinol Metab.* 2017;28(6):449–460.
45. Gill FB. *Ornithology.* 2nd ed. New York, NY: W.H. Freeman and Co; 1995.
45a. Holmes DJ, Thomson SL, Wu J, Ottinger MA. Reproductive aging in female birds. *Exp Gerontol.* 2003;38:751–756.
46. Ottinger MA, Nisbet ICT, Finch CE. Aging and reproduction: comparative endocrinology of the common tern and Japanese quail. *Am Zool.* 1995;35:299–306.
47. Nisbet ICT, Finch CE, Thompson N, Russek-Cohen E, Proudman JA, Ottinger MA. Endocrine patterns during aging in the common tern (*Sterna hirundo*). *Gen Comp Endocrinol.* 1999;114:279–286.
48. Ottinger, MA, Ricklefs, RE, Finch, CE,Ottinger MA, Ricklefs RE, Finch CE. Proceedings of the 2nd Symposium on Organisms with Slow Aging (SOSA-2). *Exp Gerontol.* 2003;38(7)
49. Holmes DJ, Ottinger MA. Biology of aging in avian species. *Am Aging Assoc J.* 2005;27:1–5.
50. Holmes DJ, Ottinger MA. Domestic and wild bird models for the study of aging. In: Conn Michael Pca, ed. In: *Handbook of Models for Human Aging. Amsterdam: Elsevier; 2006:351–366.*
51. Ottinger MA, Lavoie E. Neuroendocrine and immune characteristics of aging in avian species. *Cytogenet Genome Res.* 2007;117(1–4):352–357.
52. Robinson FE, Hardin RT, Robblee AR. Reproductive senescence in domestic fowl; effects on egg production, sequence length and inter-sequence pause length. *Br Poult Sci.* 1990;31(4):871–879.
53. Ottinger MA. Altered neuroendocrine mechanisms during reproductive aging. *Poult Sci Rev.* 1992;4:235–248.
54. Sharp PJ, Dunn IC, Cerolini S. Neuroendocrine control of reduced persistence of egg-laying in domestic hens: evidence for the development of photorefractoriness. *J Reprod Fertil.* 1992;94(1):221–235.
55. Lillpers K, Wilhelmson M. Age-dependent changes in oviposition pattern and egg production traits in the domestic hen. *Poult Sci.* 1993;72(11):2005–2011.
56. Ciccone NA, Sharp PJ, Wilson PW, Dunn IC. Changes in reproductive neuroendocrine mRNAs with decreasing ovarian function in ageing hens. *Gen Comp Endocrinol.* 2005;144 (1):20–27.
56a. Kaetzel Jr DM, Soares Jr JH. Effect of dietary calcium stress on plasma vitamin D3 metabolites in the egg-laying Japanese quail. *Poultry Sci.* 1985;64(4):1121–1127.

57. Joyner CJ, Peddie MJ, Taylor TG. The effect of age on egg production in the domestic hen. *Gen Comp Endocrinol.* 1987;65(3):331–336.
58. Ottinger MA. Aging in the avian brain: neuroendocrine considerations. *Sem Avian Ex Pet Med.* 1996;5(3):172–177.
59. Nelson BF, Daunt F, Monaghan P, Wanless S, Butler A, Heidinger BJ, Newell M, Dawson A. Protracted treatment with corticosterone reduces breeding success in a long-lived bird. *Gen Comp Endocrinol.* 2015;210:38045.
60. Delany ME, Krupkin AB, Miller MM. Organization of telomere sequences in birds: evidence for arrays of extreme length and for in vivo shortening. *Cytogenet Cell Genet.* 2000;90:139–145.
61. Aydinonat D, Penn DJ, Smith S, Moodley Y, Hoelzl F, Knauer F, Schwarzenberger F. Social isolation shortens telomeres in African Grey parrots (*Psittacus erithacus erithacus*). *PLoS One.* 2014;9(4):e93839.
61a. Haussmann MF, Heidinger BJ. Telomere dynamics may link stress exposure and ageing across generations. *Biol Lett.* 2015;11(11). doi: 10.1098/rsbl.2015.0396. pii: 20150396.
62. Downing PA, Cornawallis CK, Griffin AS. Sex, long life and the evolutionary transition to cooperative breeding in birds. *Proc Biol Sci.* 2015;282(1816):20151663.
63. Balbontin J, Moller AP. Environmental conditions during early life accelerate the rate of senescence in a short-lived passerine bird. *Ecology.* 2015;96(4):948–959.
64. Avital-Cohen N, Heiblum R, Rosenstrauch A, Chaiseha Y, Mobarkey N, Gumulka M, Rozenboim I. Role of the serotonergic axis in the reproductive failure associated with aging broiler breeder roosters. *Domest Anim Endocrinol.* 2015;53:42–51.
64a. Crews D, Willingham E, Skipper JK. Endocrine disruptors: present issues, future directions. *Q Rev Biol.* 2000;75(3):243–260.
64b. Chapman PM, Fairbrother A, Brown D. A critical evaluation of safety (uncertainty) factors for ecological risk assessment. *Environ Toxicol Chem.* 1998;17(1):99–108.
65. McLachlan JA, Korach KS. Symposium on estrogens in the environment, III. *Environ Health Perspect.* 1995;103(suppl 7):3–5.
66. Ankley G, Mihaich E, Stahl R, Tillitt D, Colborn T, McMaster S, Miller R, Bantle J, Campbell P, Denslow N, Dickerson R, Folmar L, Fry M, Giesy J, Gray E, Guiney P, Hutchinson T, Kennedy S, Kramer V, LeBlanc G, Mayes M, Nimrod A, Patino R, Peterson R, Purdy R, Ringer R, Thomas P, Touart L, Van der Kraak G, Zacharewski T. Overview of a workshop on screening methods for detecting potential (anti-) estrogenic/androgenic chemicals in wildlife. *Environ Toxicol Chem.* 1997;17(1):68–87.
67. Cheek AO, Vonier PM, Oberdorster E, Burow BC, McLachlan JA. Environmental signaling: a biological context for endocrine disruption. *Environ Health Perspect.* 1998;106(suppl 1):5–10.
68. Gray Jr LE, Ostby J, Wolf C, Lambright C, Kelce W. The value of mechanistic studies in laboratory animals for the prediction of reproductive effects in wildlife: endocrine effects on mammalian sexual differentiation. *Environ Toxicol Chem.* 1998;17:109–118.
69. NRC. *Hormonally Active Agents in the Environment.* Washington, DC, National Academy Press; 1999
70. Guillette LJ, Crain DA, Gunderson MP, Kools SAE, Milnes MR, Orlando EF, Rooney AA, Woodward AR. Alligators and endocrine disrupting contaminants: a current perspective. *Amer Zool.* 2000;40:438–452.
71. Ryberg KR, Gilliom RF. Trends in pesticide concentrations and use for major rivers of the United States. *Sci Total Environ.* 2015;538:431–444.
72. Heindel JJ, Newbold R, Schug TT. Endocrine disruptors and obesity. *Nat Rev Endocrinol.* 2015;11(11):653–661.
73. Janesick AS, Blumberg B. Obesogens: an emerging threat to public health. *Am J Obstet Gyecol.* 2016;214(5):559–565.

74. Heindel JJ, Blumberg B, Cave M, Machtinger R, Mantovani A, Mendez MA, Nadal A, Palanza P, Panzica G, Sargis R, Vandenberg LN, Vom Saal F. Metabolism disrupting chemicals and metabolic disorders. *Reprod Toxicol.* 2017;68:3–33.
74a. Polyzos SA, Kountouras J, Deretzi G, Zavos C, Mantzoros CS. The emerging role of endocrine disruptors in pathogenesis of insulin resistance: a concept implicating nonalcoholic fatty liver disease. *Curr Mol Med.* 2012;12(1):68–82.
75. Chevalier N, Fenichel P. Endocrine disruptors: new players in the pathophysiology of type 2 diabetes? *Diabetes Metab.* 2015;41(2):107–115.
76. Nalbone G, Cicolella A, Laot-Cabon S. Endocrine disruptors and metabolic diseases: a major public health challenge. *Reprod Toxicol.* 2013;25(1):45–49.
77. Foulds CE, Trevino LS, York B, Walker CL. Endocrine-disrupting chemicals and fatty liver disease. *Nat Rev Endocrinol.* 2017;13(8):445–457.
78. Chamorrow-Garcia R, Blumberg B. Transgenerational effects of obesogens and the obesity epidemic. *Opin Pharmacol.* 2014;19:153–158.

FURTHER READING

Crews D, McLachlan JA. Epigenetics, evolution, endocrine disruption, health, and disease. *Endocrinology.* 2006;147(6):S4–S10.

Fairbrother A. Comparative aspects of estrogen functions and measurements in oviparous and vivparous vertebrates. *Hum Ecol Risk Assess.* 2000;6(1):73–102.

Finch CE. Update on slow aging and negligible senescence—a mini-review. *Gerontology.* 2009;55(3):307–313.

Finch CE, Crimmins EM. Constant molecular aging rates vs. the exponential acceleration of mortality. *Proc Natl Acad Sci USA.* 2016;113(5):1121–1123.

Hester PY, Enneking SA, Haley BK, Cheng HW, Einstein ME, Rubin DA. The effect of perch availability during pullet rearing and egg laying on musculoskeletal health of caged White Leghorn hens. *Poult Sci.* 2013;92(8):1972–1980.

Kaetzel Jr DM, Soares Jr JH. Effect of dietary calcium stress on plasma vitamin D3 metabolites in the egg-laying Japanese quail. *Poultry Sci.* 1985;64(4):1121–1127.

Mair W, Sgra CM, Johnson AP, Chapman T, Partridge L. Lifespan extension by dietary restriction in female *Drosophila melanogaster* is not caused by a reduction in vitellogenesis or ovarian activity. *Exp Gerontol.* 2004;39:1011–1019.

Merry BJ. Molecular mechanisms linking calorie restriction and longevity. *Int J Biochem Cell Biol.* 2002;34:1340–1354.

Montgomery MK, Buttemer WA, Hulbert AJ. Does the oxidative stress theory of aging explain longevity differences in birds? II. Antioxidant systems and oxidative damage. *Exp Gerontol.* 2012;47(3):211–222.

Ottinger MA, Mobarak M, Abdelnabi MA, Roth G, Proudman JA, Ingram DK. Effects of caloric restriction on reproductive and adrenal systems in Japanese quail: are responses similar to mammals, particularly primates? *Mech Ageing Dev.* 2005;126(9):967–975.

Pasquali R. Obesity and androgens: facts and perspectives. *Fertil Steril.* 2006;85(5):1319–1340.

Pasquali R, Vicennati V, Pasquali R, Vicennati V. Steroids and the metabolic syndrome. *J Steroid Biochem Mol Biol.* 2008;109(3–5):258–265.

Polyzos SA, Kountouras J, Deretzi G, Zavos C, Mantzoros CS. The emerging role of endocrine disruptors in pathogenesis of insulin resistance: a concept implicating nonalcoholic fatty liver disease. *Curr Mol Med.* 2012;12(1):68–82.

Roth GS, Lane MA, Ingram DK, Mattison JA, Elahi D, Tobin JD, Muller D, Metter EJ. Biomarkers of caloric restriction may predict longevity in humans. *Science.* 2002;297(5582):811–814.

INDEX

B

C

D

E

F

MIX

Papier | Fördert gute Waldnutzung

FSC® C083411

Kolibri 360 GmbH
Ferdinand-Jühlke-Straße 7
99095 Erfurt, Deutschland
produktsicherheit@kolibri360.de